375
14

PRENTICE-HALL FOUNDATIONS OF MODERN BIOLOGY SERIES

William D. McElroy and Carl P. Swanson, Editors

NEW VOLUME

Chemical Background for the Biological Sciences, Emil H. White

SECOND EDITIONS

The Cell, Carl P. Swanson

Cell Physiology and Biochemistry, William D. McElroy

Heredity, David M. Bonner and Stanley E. Mills

Adaptation, Bruce Wallace and Adrian M. Srb

Growth and Development, Maurice Sussman

Animal Physiology, Knut Schmidt-Nielsen

Animal Diversity, Earl D. Hanson

Animal Behavior, V. G. Dethier and Eliot Stellar

The Life of the Green Plant, Arthur W. Galston

The Plant Kingdom, Harold C. Bold

Man in Nature, Marston Bates

DAVID M. BONNER *University of California, San Diego*

STANLEY E. MILLS *University of California, San Diego*

Englewood Cliffs, N. J. **PRENTICE-HALL, INC.**

Heredity

SECOND EDITION

FOUNDATIONS OF MODERN BIOLOGY SERIES

48,880

HEREDITY, SECOND EDITION, *David M. Bonner and Stanley E. Mills*

© *Copyright 1961, 1964 by* PRENTICE-HALL, INC.
Englewood Cliffs, New Jersey
All rights reserved. No part of this book
may be reproduced in any form, by mimeograph or any other means,
without permission from the publishers.
Library of Congress Catalog Card Number: 64–15786
Printed in the United States of America.

FOUNDATIONS OF MODERN BIOLOGY SERIES
William D. McElroy and Carl P. Swanson, Editors

Design by Walter Behnke

Drawings by Felix Cooper

PRENTICE-HALL INTERNATIONAL, INC., *London*
PRENTICE-HALL OF AUSTRALIA, PTY., LTD., *Sydney*
PRENTICE-HALL OF CANADA, LTD., *Toronto*
PRENTICE-HALL OF INDIA PVT. LTD., *New Delhi*
PRENTICE-HALL OF JAPAN, INC., *Tokyo*
PRENTICE-HALL DE MEXICO, S. A., *Mexico City*

C-38675 (p)   *C-38676 (c)*

Foundations
of Modern Biology
Series

The science of biology today is *not* the same science of fifty, twenty-five, or even ten years ago. Today's accelerated pace of research, aided by new instruments, techniques, and points of view, imparts to biology a rapidly changing character as discoveries pile one on top of the other. All of us are aware, however, that each new and important discovery is not just a mere addition to our knowledge; it also throws our established beliefs into question, and forces us constantly to reappraise and often to reshape the foundations upon which biology rests. An adequate presentation of the dynamic state of modern biology is, therefore, a formidable task and a challenge worthy of our best teachers.

The authors of this series believe that a new approach to the organization of the subject matter of biology is urgently needed to meet this challenge, an approach that introduces the student to biology as a growing, active science, and that also *permits each teacher of biology to determine the level and structure of his own course.* A single textbook cannot provide such flexibility, and it is the authors' strong conviction that these student needs and teacher prerogatives can

v

best be met by a series of short, inexpensive, well-written, and well-illustrated books so planned as to encompass those areas of study central to an understanding of the content, state, and direction of modern biology. The FOUNDATIONS OF MODERN BIOLOGY SERIES represents the translation of these ideas into print, with each volume being complete in itself yet at the same time serving as an integral part of the series as a whole.

PREFACE TO THE SECOND EDITION

The first edition of the FOUNDATIONS OF MODERN BIOLOGY SERIES represented a marked departure from the traditions of textbook writing. The enthusiastic acceptance of the Series by teachers of biology, here and abroad, has been most heartening, and confirms our belief that there was a long-felt need for flexible teaching units based on current views and concepts. The second edition of all volumes in the Series retains the earlier flexibility, eliminates certain unnecessary overlaps of content, introduces new and relevant information, and provides more meaningful illustrative material.

The Series has also been strengthened by the inclusion of a new volume, *Chemical Background for the Biological Sciences* by Dr. Emil White. The dependence of modern biology on a sound foundation in physics and chemistry is obvious; this volume is designed to provide the necessary background in these areas.

In preparing the second edition of the Series, the authors and editors gratefully acknowledge the many constructive criticisms that have been made by hundreds of teaching biologists. Their interest and aid have made the task of writing more a pleasure than a burden.

CURIOSITY

may have killed the cat; more likely
the cat was just unlucky, or else curious
to see what death was like, having no cause
to go on licking paws, or fathering
litter on litter of kittens, predictably.
    Nevertheless, to be curious
is dangerous enough. To distrust
what is always said, what seems,
to ask odd questions, interfere in dreams,
leave home, smell rats, have hunches
does not endear him to those doggy circles
where well-smelt baskets, suitable wives, good lunches
are the order of things, and where prevails
much wagging of incurious heads and tails.
    Face it. Curiosity
will not cause him to die—

only lack of it will.
Never to want to see
the other side of the hill,
or that improbable country
where living is an idyll
(although a probable hell)
would kill us all.
Only the curious
have, if they live, a tale
worth telling at all.

Dogs say he loves too much, is irresponsible,
is changeable, marries too many wives,
deserts his children, chills all dinner tables
with tales of his nine lives.
Well, he is lucky. Let him be
nine-lived and contradictory,
curious enough to change, prepared to pay
the cat price, which is to die
and die again and again,
each time with no less pain.
A cat minority of one
is all that can be counted on
to tell the truth. And what he has to tell
on each return from hell
is this: that dying is what the living do,
that dying is what the loving do,
and that dead dogs are those who do not know
that hell is where, to live, they have to go.

ALASTAIR REID

© 1959 The New Yorker Magazine, Inc.

Contents

ix

HEREDITY

The Material
Basis of Heredity

The word heredity is simply the name given to the means by which living organisms reproduce their kind. An understanding of heredity, therefore, requires knowledge of the various processes involved in reproduction. In recent years, this knowledge has increased dramatically. The combined efforts of an ever increasing number of research scientists, and the development of new tools and experimental techniques of great power, have already revealed much, and promise more. Many questions that have puzzled generations of biologists have been answered. More important, however, new information has bred new questions, provoked a desire for deeper levels of comprehension. Today, with the descriptive elements of reproduction largely clarified and behind us, we are going after the very atoms of heredity. We want to know about their arrangements, their architecture and their activities, the chemical transformations and the physical forces that provide for the continuation of life.

We can begin by noting a fact obvious to all of us. Living systems are highly complex. For example, the most elementary studies soon reveal the wealth of biochemical activities that a cell must contain just to survive. Yet we also know from everyday observation that each organism must have not only the means to maintain itself, but information for reproducing systems like itself as well. The best evidence available today indicates that this information is contained—coded is the popular term—in a remarkable set of substances which is passed on from one generation to the next, and which in large measure prescribes the nature of the succeeding generation. These substances we refer to as "the material basis of heredity."

The scientific study of the "material basis" of heredity began in the latter part of the nineteenth century with the work of the Austrian monk, Gregor Mendel. His experimental results led him to the formulation of the "laws of heredity," which, in turn, provided the inspiration for the research that continues at an accelerated pace today. Of great current interest is the fact that recent work in this field has brought about a revolution in modern experimental biology. It has led to the development of what is now known as molecular biology, in which the disciplines of biology, chemistry, and physics have fused to create a field of challenge and excitement. It is this field that shows explosive experimental activity at the present time, and it is the authors' bias that molecular biology offers the best opportunity for gaining insight into the basic problems of biology as they are now defined. Hence, the present discussion of heredity will be centered around the developments of recent years, developments using as experimental tools microorganisms rather than higher plants or higher animals. Our hope is that in this way it will be possible not only to present clearly the principles underlying the science of heredity, i.e., genetics, but at the same time to give some insight into the problems of the present and the fascination of the future.

THE CELLULAR ELEMENTS

All organisms consist of cells. Moreover, whether they consist of one cell or of thousands, they grow and reproduce by cell division. A bacterium is a single cell. It grows until it has doubled in size, then reproduces by splitting in two. The result of this division: two identical cells from one. Man is a multicellular organism and he, too, grows by increase in cell size and division of his constituent cells. He reproduces by producing certain cells (gametes), sperm (male) and eggs (female), which fuse to give rise to a single fertilized egg cell (zygote). The zygote is capable of growing and dividing, and of ultimately giving rise to a new multicellular individual. The information needed to reproduce a complex multicellular individual is thus transmitted through a single cell, and each dividing cell in turn must contain the information required to reproduce itself. The elements of heredity belong to the cell, therefore, rather than to the organism, and, in considering the material basis of heredity, we are free to focus on the processes involved in the reproduction of single cells.

What elements within cells can be expected to carry hereditary information? Consider the structure of a cell. The basic structures of most cells are remarkably similar (Fig. 1-1). A cell is surrounded by a cell membrane, and sometimes an outer cell wall. Within the membrane there is the nucleus and cytoplasm. The nucleus of most cells is a discrete body, having a nuclear membrane and containing thread-like structures (chromosomes).

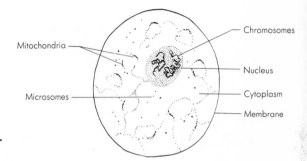

Fig. 1-1. Diagrammatic representation of a cell.

The cytoplasm contains a number of discrete elements. Mitochondria (small particles having a defined and characteristic morphology) are found in profusion in the cytoplasm, and, by means of an electron miscroscope, we can see still smaller particles (microsomes). Thus, the cytoplasm is found to contain different kinds of particulate elements, varying in number and, as will be seen later, in function, whereas in general but a single nucleus is found per cell.

The cellular elements that carry hereditary information must be present in every reproducing cell. In addition, to account for the fact that like faithfully begets like, these elements must be capable of reproducing or dividing accurately. The nucleus fulfills these requirements. For example, every cell capable of reproduction has a nucleus; cells lacking one cannot reproduce. Not every cell has a well-defined nucleus—the bacteria, for example—but even they have characteristic nuclear bodies. During cell division, the nucleus divides by an elaborate mechanism (mitosis) that leads to the production of an exact copy of each constituent chromosome. It is the chromosomes in the nucleus that carry the material basis of heredity. A picture of a virus chromosome is shown in Fig. 1-2.

The preceding paragraph, so easily written, sums up about half a century of unrelenting research. For a glimpse of the kind of evidence leading to such economical, heroic prose, let us consider briefly a few of the experiments of Gregor Mendel and Thomas Hunt Morgan. Mendel, working with the common garden pea, patiently collected many varieties

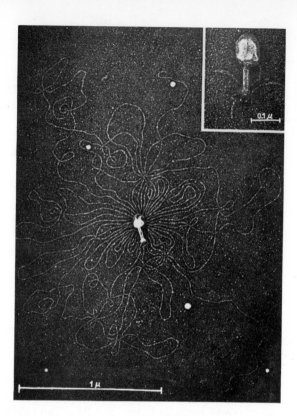

Fig. 1-2. An electron micrograph of the chromosome of a bacterial virus, T_2. Two free ends are visible. The inset shows the organism from which the chromosome was derived and illustrates the problem of packaging. (Kleinschmidt, Lang, Jacherts and Zahn. Biochimica et Biophysica Acta 61 (1962), pp. 857–864.)

that were true-breeding for certain characteristics—e.g., flower color, seed-coat texture. True-breeding means that a red variety, for example, when self-fertilized with its own pollen, yields only red progeny. Similarly, white gives white. When Mendel crossed the red and white varieties, he found the progeny (first filial generation, F_1) to be all red. Question: What had happened to the trait for white color? The answer came when he self-fertilized the F_1 red plants to obtain the second generation, F_2. About one-fourth of the F_2 were white; they were all *true-breeding* and indistinguishable in color from the original white variety. Whatever it was, then, that determined the white color had survived intact its submergence in the F_1 generation and was now doing fine. A similar analysis of a cross between plants true-breeding for either wrinkled or smooth seed covers gave the same result. The wrinkled trait vanished in the F_1 generation only to reappear uncorrupted in F_2. Mendel then analyzed the simultaneous inheritance in F_1 and F_2 of both traits (quantitative details will come later). He found that the inheritance of smooth or wrinkled seeds was uninfluenced by the inheritance of red or white flower color. Extending his analysis to other pairs of traits, Mendel's observations led him to conclude that in sexual reproduction the inheritance of one trait occurred independently of the inheritance of another.

Many years later, however, Morgan and his associates, studying the fruit fly, found that certain combinations of traits were transmitted together, suggesting that the hereditary elements determining these traits were, in some way, linked. Analysis of the inheritance of many traits showed that they could be separated into several groups. Within each group, the traits, and therefore the hereditary elements controlling them, were inherited together, i.e., they were linked. On the other hand, the inheritance of traits, each of which fell into a different group, showed the independence observed by Mendel. In *Drosophila melanogaster,* four and only four linkage groups could be found. Cytological observations revealed that the cells of Drosophila contain precisely four distinct pairs of chromosomes. An interesting coincidence.

Morgan thereupon concluded that the chromosomes must be the hereditary elements, and he was right. This can be proven by cytological study of the giant chromosomes present in the cells of the salivary glands in Drosophila. For example, a fly was found to have an abnormal trait, and this was reflected in an abnormality in the structure of one of its chromosomes. When this fly was mated to a normal fly and the progeny examined, it developed that every fly with the abnormal chromosome had the abnormal trait. All the normal progeny had the normal chromosome. Thus the combined use of cytology and genetic analysis led to the clear proof that the chromosomes of the nucleus carry the genetic material.

The number of chromosomes per nucleus is characteristic and constant for each living organism. The somatic cells of higher plants and animals contain duplicate sets of chromosomes (diploid $= 2n$), while the gametic cells contain a single set (haploid $= n$). The gametic cells of Drosophila have a chromosome number of 4, the somatic cells have 8. The haploid number for human cells is 23, for the fungus *Neurospora crassa* 7, and for the bacterium *E. coli* 1. There is a clear species specificity with respect to chromosome number and chromosome morphology.

MITOSIS

Since chromosomes carry the hereditary information, we would expect them to be accurately reproduced and transmitted in an orderly and precise way. They are. Sometime during the life history of a cell, in the time between cell divisions, an exact copy of each chromosome is synthesized. When cell division is triggered, a sequence of events occurs which results in the formation of two cells with identical sets of chromosomes. This type of cell and nuclear division is called mitosis. We don't know the intimate details of this division for every cell. Bacterial cells are too small for mitosis to be seen in the light microscope; but in cells of higher organisms we can watch and describe the mitotic sequence. At first glance, the various stages of mitosis appear overwhelmingly complicated, but what is

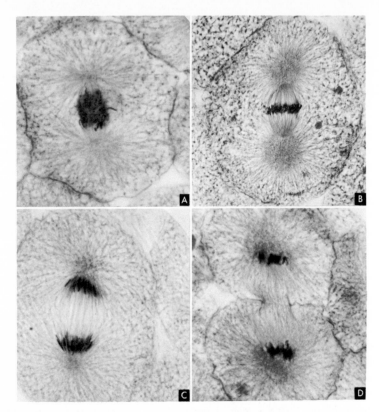

Fig. 1-3. Photograph of mitosis in the whitefish. When cell division is triggered:
(A) The chromosomes coil up and gather, each with its newly synthesized and
still attached duplicate copy (sister chromatids). (B) Alignment before separation
of sister chromatids. (C) Separation. (D) Formation of two new cells with identical
chromosome sets. (Copyright by General Biological Supply House, Inc., Chicago.)

Fig. 1-4. Diagrammatic representation of mitosis for a haploid (n = 2) and a
diploid (2n = 4) cell. (A) Original cells. (B) Chromosomes lined up in a plane at
random. (C) Two progeny cells with chromosome content identical to original
cell. Different chromosomes are labelled I and II and homologous pairs as Iᵃ,
Iᵇ, and IIᵃ, IIᵇ.

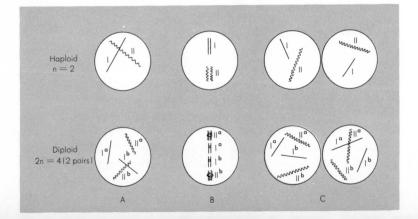

accomplished is simple. In brief, the chromosomes contract and gather independently of one another on a plane approximately in the center of the cell. Each chromosome is then seen to be tightly paired with its previously synthesized copy. The sister strands, or chromatids, then separate, one to each end of the cell, and two identical sets of chromosomes are formed. The nuclei reorganize, and the cell divides in two. A mitotic cell division is simply a division that produces two cells, each with identical chromosomes (Fig. 1-3). A diagram of mitosis for a haploid cell ($n = 2$) and for a diploid cell ($2n = 4$) is shown in Fig. 1-4.*

MEIOSIS

Both haploid and diploid cells can undergo mitosis. In fact, mitosis or an event closely similar to mitosis is the general mechanism for *nonsexual* reproduction of all living cells. There is another general mechanism for the distribution of chromosomes from one cell to its daughter cells. This mechanism, called *meiosis,* is invoked in the case of sexual reproduction in which two cells fuse, each cell bringing to the union its own complement of chromosomes. We postpone a description of meiosis to Chapter Three, but it is pertinent to observe here that both mitosis and meiosis provide for an orderly distribution of chromosomes from a cell to its progeny cells. As will be made clear in subsequent discussion, the laws of heredity reflect the characteristics of this process. However, a clear understanding of heredity requires knowledge of chromosomes beyond that concerned with the mechanisms of transmission and distribution. A crucial bit of information is, how do chromosomes dictate specific traits? By what chemical and physical legerdemain does a chromosome determine whether a person is male or female, his skin black or white, his wit quick or nonexistent? This is a problem of molecular genetics, for here we must define the chemical nature of the hereditary material contained within the chromosome itself and, of equal importance, the chemical reactions this material carries out that enable it to specify cellular characteristics. Answers to these problems are essential to our understanding of living matter, and it is this aspect of genetics, molecular genetics, that will be initially discussed.

* A well-written and beautifully illustrated account of cell division can be found in an article by Daniel Mazia in the *Scientific American,* September, 1961.

DNA:
The Genetic
Material

CHAPTER TWO

Since both cytological and genetic evidence give overwhelming proof that chromosomes carry the hereditary material, a study of the chemical nature of chromosomes might be expected to give insight into the chemical basis of heredity. Within recent years, it has been possible to isolate chromosomes from the other constituents of the cell and do chemical analysis. It turns out that chromosomes are complex, not only in terms of morphology, but in chemical terms as well. They do not consist of a single chemical substance, but are composed, chiefly, of three types of substances: proteins, deoxyribonucleic acid (DNA), and ribonucleic acid (RNA).

A brief word about these three substances. They are naturally occurring polymers. A polymer is a large molecule formed from a few simple molecules (the "mers") repeatedly linked in chemical bondage. Proteins are made up of various combinations of 20 simple molecules—amino acids—and different proteins may have anywhere from 100 to 10,000 units. DNA and RNA are generally made by repeated linkage of four simple units called nucleotides. A DNA molecule may have as many as 100,000 units strung together. The chemical difference between DNA and RNA, the significance of the order in which the units are linked (primary structure), and the total number of each unit per molecule (composition) will be discussed later. It is sufficient for now to note that from pure chemical analysis of chromosomes, we are not able to conclude whether one of these three substances contains the genetic information, or whether it is contained in some combination. This is perhaps not too surprising. Proof of this point must come ultimately from biological tests.

In order to establish that a certain substance contains genetic information, we would like to isolate the substance in pure form from one organism and demonstrate that when it is put into another organism, traits of the first organism appear in the second and are passed on to the progeny of the second—a simple and sensitive test. We do precisely this in the next paragraph.

TRANSFORMATION

The lines of inquiry that led to an understanding of the chemical nature of genetic material arose from a study of the pestilent organism *Diplococcus pneumoniae*. This bacterium, when in a "virulent" state, causes pneumonia in man and a fatal bacterial infection on injection into mice. Certain strains of this organism are "avirulent" in that they do not cause disease. Both virulence and avirulence are stable, heritable properties in that they are passed on from the bacterium to its progeny upon cell division.

During the 1920's two bacteriologists, working with the pneumococcus, confirmed that one of their strains was avirulent. It produced no disease in mice. In addition, if they injected mice with a virulent strain previously killed by heat, again there were no symptoms (Fig. 2-1). In the crucial experiment mice were inoculated with both living, avirulent bacteria and heat-killed bacteria. Even though neither alone was sufficient for disease production, symptoms of disease did appear, and, upon isolation of the organism giving rise to these symptoms, a virulent strain was found. Moreover, the disease-producing trait was heritable, the strain retaining its virulence through countless cell divisions. The presence of heat-killed virulent cells, therefore, had the curious effect of changing avirulent cells to virulent cells, which is a change in a genetic trait. This phenomenon was called bacterial transformation, and its study by Oswald Avery and his collaborators at the Rockefeller Institute opened the way to the identification of the chemical nature of the genetic material.

It is now known that many inherited bacterial traits will undergo transformation, and that transformation can be carried out in a test tube as well as in mice. For instance, again with the pneumococcus organism as an example, certain strains are killed by the antibiotic streptomycin (and are thus called streptomycin-sensitive). This sensitivity is passed on from one

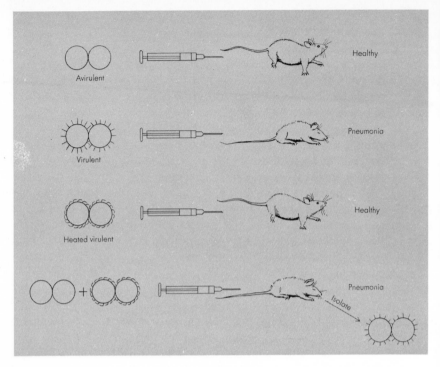

Fig. 2-1. Bacterial transformation in living animals in vivo.

cell to another at cell division. Other strains are known to be resistant to streptomycin, also a heritable trait. Suppose we perform the following experiment (see Fig. 2-2). Grow a large mass of streptomycin-resistant bacteria, kill them, then make a dilute salt extract. Add some extract of the resistant strain to a growing population of a *sensitive* strain. Many cells of the sensitive strain are now converted to streptomycin-resistant cells, and they pass this resistance on to their progeny. Extracts of streptomycin-resistant cells, therefore, have the effect of causing a permanent hereditary change from streptomycin sensitivity to streptomycin resistance.

It then proved possible, by chemical fractionation, to isolate the substance in such extracts that causes this change in genetic traits. It was shown that purified DNA present in such extracts is the active component. This extraordinary phenomenon, molecular sex if you will, may not captivate you from an esthetic point of view, but its importance in terms of its contribution to the understanding of genetic material cannot be overestimated.

DNA has been shown to convey heritable traits from one bacterium to another in at least six different strains of bacteria. Transformable traits include such characteristics as ability to produce disease, antibiotic resistance, and nutritional requirements. Very recently, it has been shown that bacteria, essentially autonomous single-celled organisms, enjoy no monopoly on transformation. Among the many things being done by biologists today, methods have been developed that make it possible to isolate single cells from metazoans like ourselves, and grow large populations from these animal cells in the laboratory. These operations are part of a field called tissue culture. Elizabeth and Waclaw Szybalski, working in tissue culture with human cells derived from bone marrow, obtained two cell lines. One of these lines synthesized an enzyme called inosinic acid pyrophosphorylase; the other cell line didn't. (You will be reading more, much more, about enzymes and what they do.) The capacity to make an enzyme is a heritable trait and the Szybalskis, using purified human DNA from the possessor cell line, were able to transform cells of the deficient cell lines to enzyme production, and the gift was visited on all their progeny.

DNA thus fulfills the criteria set forth earlier, i.e., for a certain chemical substance to be designated as containing the genetic information of a given organism, it should be present in the nucleus or nuclear bodies and we should be able to isolate the substance and demonstrate that when it is put into another organism, traits of the donating organism appear in the second. Neither RNA nor protein are active in bacterial transformation. Thus we may conclude that DNA is the chemical substance which serves as the basis of heredity.

BACTERIAL VIRUSES

A second line of inquiry that has also led to identification of the genetic material as DNA is the study of bacterial viruses. For a better understanding of the nature of the evidence, it would be well to describe a typical bacterium and bacterial virus.

Bacteria are microscopic, single-celled organisms. They have characteristic nuclear-like bodies (nucleoids) which contain DNA. Bacterial cells also contain large amounts of RNA, among other substances. Bacterial

Fig. 2-2. Bacterial transformation in the test tube (in vitro), **using purified DNA and the trait of streptomycin resistance.**

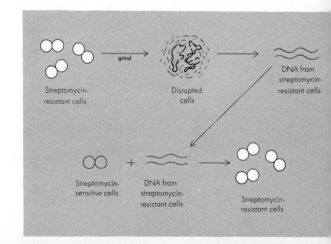

viruses are still smaller, but they can be seen by means of an electron microscope (Fig. 2-3). Bacterial viruses cannot reproduce in the absence of living bacterial cells, and most are found to consist of protein, DNA, and a few small molecules.

The life cycle of a typical bacterial virus (bacteriophage or just phage) is given in Fig. 2-4. The life cycle may be divided into three distinct phases: the infective phase, the vegetative phase, and the progeny-formation phase. The infective particle attaches to the bacterial cell wall by its tail. A break in the cell wall is induced, and the contents of the infecting particle are then injected into the bacterial cell. This step is of great interest, because the injecting particle has a protein coat around a core consisting mainly of DNA. Thus the material that is injected into the bacterial cell and that initiates the formation of new virus particles is DNA.

After injection of the virus DNA, the bacterial cell goes into a period of vegetative virus formation during which new virus DNA and protein are synthesized inside the bacterium. In the final stage of this 20-minute life cycle, the newly synthesized virus protein and DNA are assembled into new virus particles, the cell ruptures, and the viruses spill out into the medium. The point of particular interest here is that the material which is injected into the bacterial cell, and which initiates the formation of new virus particles, is DNA. The genetic information of bacterial viruses is therefore contained in DNA. Study of both bacterial transformation and bacterial viruses consequently leads to the conclusion that DNA is the chemical basis of heredity.

Fig. 2-3. Electron micrograph of a bacterial virus. (Courtesy Dr. L. W. Labaw.)

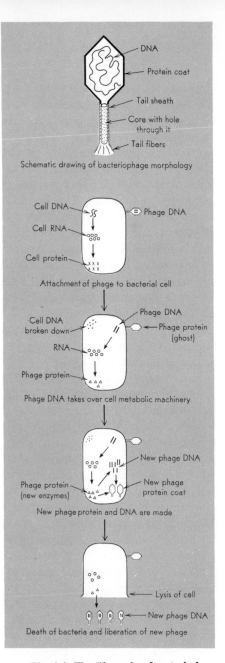

DNA

Protein coat

Tail sheath

Core with hole through it

Tail fibers

Schematic drawing of bacteriophage morphology

Cell DNA

Phage DNA

Cell RNA

Cell protein

Attachment of phage to bacterial cell

Cell DNA broken down

Phage DNA

Phage protein (ghost)

RNA

Phage protein

Phage DNA takes over cell metabolic machinery

New phage DNA

Phage protein (new enzymes)

New phage protein coat

New phage protein and DNA are made

Lysis of cell

New phage DNA

Death of bacteria and liberation of new phage

Fig. 2-4. The life cycle of a typical bacterial virus or bacteriophage. The infective phase consists of attachment and injection. The vegetative phase is the period required for the formation of new phage protein and DNA. Progeny formation requires the assembly of phage DNA and protein into new infective particles.

This conclusion, however, brings up a contemporary problem of great importance. Many different viruses are, of course, known. They are all extraordinarily small in size, and they all have an obligate requirement for a living host for replication. In addition to bacterial viruses, plant and animal viruses are well known. Although bacterial viruses consist of protein and DNA, many plant and animal viruses are known that consist of protein and RNA instead. Tobacco mosaic virus, a well-known virus of tobacco plants, is of this latter type, as is the virus that causes human poliomyelitis. Using the tobacco mosaic virus (TMV), we can carry out some very interesting experiments. By chemical means, it is possible to separate the protein and RNA components of TMV. (These can be put back together in the test tube, where some stable viruses are reformed; thus at least some of the protein and RNA survive the separation uninjured.) If we inoculate a tobacco plant with just the protein fraction, nothing happens. If, however, just the RNA component is inoculated, new TMV particles are produced that have both the protein and the RNA components. The TMV RNA, therefore, carries all the information necessary for the synthesis of both the TMV protein and the RNA.

In view of such observations, how sure are we that, in higher organisms, the genetic information is coded in the DNA of the chromosomes? Very! The question that concerns us is no longer whether DNA is sufficient for genetic information—it is. Rather, is it necessary? Are there any other molecules that, under special conditions, can substitute for DNA in this function? Apparently, if tobacco mosiac virus will not be denied, certain kinds of RNA can do this. But we know relatively little about the role of this kind of RNA in the cells of higher plants and animals. Despite broad similarities in chemical composition and constitution between RNA and DNA, each seems to play a subtly different role in the cell's economy. Perhaps the important fact to know here is that DNA is the only substance which, if transferred from one *cell* to a second *cell,* results in hereditary changes in the second *cell* (transformation), and thus gives a strong indication that it is the primary substance in which genetic information is coded. The *translation* of genetic information into action involves, as we will see, RNA. These two polymers are intimately related, and the very closeness of this relationship could well result in RNA's exhibiting an apparent primary genetic function under certain circumstances, as in TMV.

At the present time, therefore, it seems reasonable to conclude that in organisms, in general, the genetic information is coded in the DNA of the chromosomes. This is reason enough to look at it.

THE STRUCTURE OF DNA

DNA has been isolated from a wide variety of organisms and, independent of the source, is found to have many of the same chemical and physical properties. DNA is a macromolecule with molecular weight ranging from 10,000 to 30,000,000 depending on the source of isolation. In general, it is composed of four small repeating units called nucleotides. The nucleotides are fairly complicated molecules themselves. They all contain phosphoric acid and the 5-carbon sugar, deoxyribose. Where they differ is in their third component, the base. There are four bases most frequently found in DNA, two purines (adenine, guanine) and two pyrimidines (cytosine, thymine). The individual bases in conjunction with deoxyribose and phosphoric acid give four nucleotides: the purine nucleotides, deoxyadenylic and deoxyguanilic acids, and the pyrimidine nucleotides, deoxythymidilic and deoxycytidylic acids. The chemical structures of the bases, and one purine nucleotide are given in Fig. 2-5. Included in the figure is the bond that links one unit in the chain with another. This is a phosphoester bond which links the deoxyribose (carbon 3) of one nucleotide to the deoxyribose (carbon 5) of its neighbor. The DNA chain is made up of a deoxyribose-phosphate backbone with the bases pointing inward and perpendicular to the fiber axis.

A number of years ago, Irwin Chargaff and his collaborators extracted

DNA from a variety of sources and measured the relative proportions of the four nucleotides in the polymer, i.e., the base composition. Some interesting results turned up. With any *given nucleotide* (take adenine), the relative amount of this unit in the DNA from, say, animal, insect, plant, fungal, and bacterial cells, varied tremendously. However, the astonishing thing was that with any *given DNA,* if all the purine nucleotides were added up and all the pyrimidine nucleotides added up, the two sums were equal. In addition, it was noted that the number of deoxyadenylic acid molecules was always equal to the number of deoxythymidylic acid molecules. Similarly, deoxyguanylic acid was equal to deoxycytidylic acid. The following is an artificial but illustrative example:

	Deoxythymidylic acid	*Deoxycytidylic acid*	*Deoxyadenylic acid*	*Deoxyguanylic acid*
	(Amount of each unit in percentages)			
DNA$_1$	10	40	10	40
DNA$_2$	30	20	30	20

THE WATSON-CRICK MODEL

These observations, together with studies of X-ray diffraction patterns by M. H. F. Wilkins, led James Watson and Francis Crick to propose in 1954

Fig. 2-5. The bases of DNA, and 5-deoxyadenylic acid (2′-deoxyadenosine-5′-phosphate).

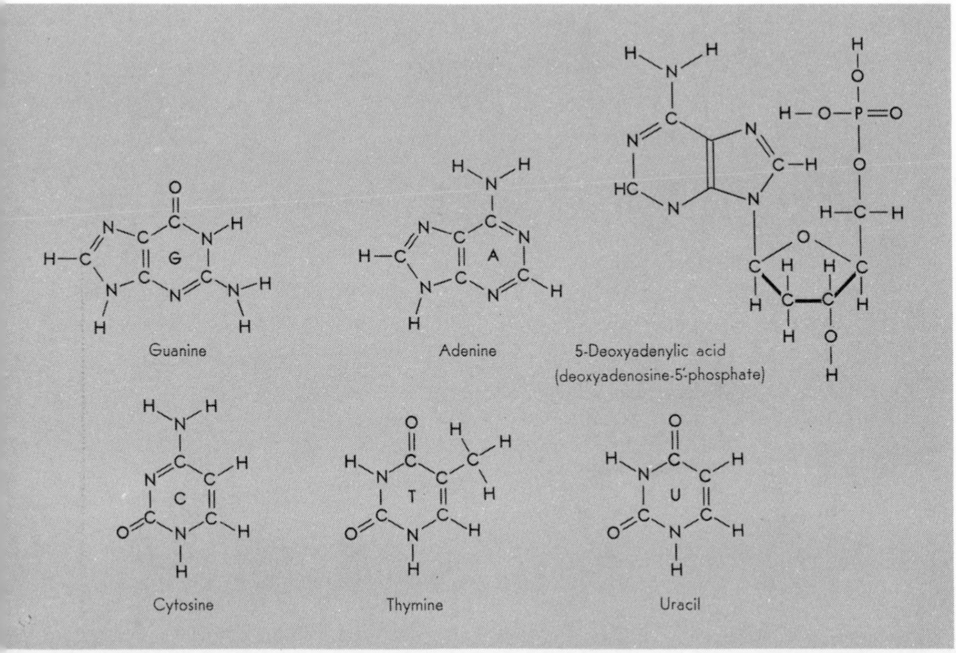

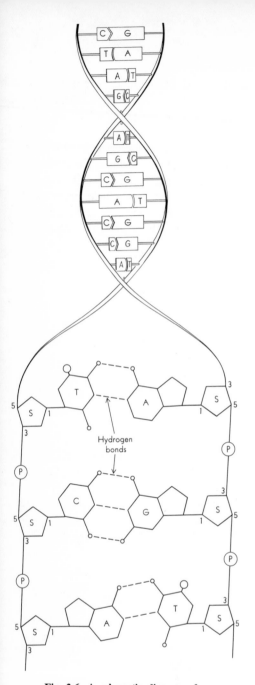

a structure for DNA which is shown in Fig. 2-6. In 1962 these three scientists were awarded the Nobel Prize for their brilliant contributions. Watson and Crick suggested that each molecule of DNA is made up of two intertwined strands, and that the sequence of bases in one strand determines the sequence of the other, the complementary strand, as a consequence of pairing through hydrogen bonding between the purine and pyrimidine bases. Wherever adenine occurs in one strand, pairing in the complementary strand is with thymine; guanine is paired with cytosine. The outstanding characteristics of the proposed structure of DNA are that it is double-stranded and that the two strands are complementary. Extensive physical and chemical investigations since 1954 have confirmed the double-stranded picture of DNA, and our theories on the mode of replication and mechanism of gene action must now include these facts. Armed with the knowledge that the material basis of heredity is DNA, let us next consider how this substance acts.

Fig. 2-6. A schematic diagram of the Watson-Crick model of the DNA molecule showing the base pairing and the helical structure.

Genes
and Biochemical
Reactions

Long before the discovery of DNA as the chemical basis of heredity, the inheritance of well-defined characteristics in insects and domestic plants had been studied by means of the basic tool of genetics, the inheritance test. The inheritance test consists of controlled matings and careful observation of the resulting progeny. For example, two plants which differ in a given trait such as color are mated (crossed), and the distribution of color in their offspring and the descendants of their offspring is then analyzed. This test is as old as genetics, and was the experimental tool that enabled Mendel to formulate his laws of inheritance.

Detailed study of the transmission of specific traits reveals that they appear to be determined by specific areas of a chromosome, called genes.

Since the chemical material in the chromosome that controls traits is DNA, a gene is really a unit of DNA. Genes exist in alternative forms and give rise in an organism to alternative versions of a given trait (if this were not true, the inheritance test would be impossible). The alternative forms of a single gene are called alleles. For example, one aspect of

17

the property of blood-clotting in man is determined by a specific region in one of his chromosomes. The possession of this region in one allelic form insures normal blood-clotting after injury. On the other hand, a carrier of an alternative form of this very same chromosomal region suffers from a disease known as hemophilia, in which blood-clotting occurs very slowly if at all. Certain genes are known to have as many as 100 different alleles, and as we become more clever in our prying, we will probably find that all genes exist in many alternate forms.

How are the genes assembled? Through patient and meticulous study of the transmission of specific traits, individually and in combination, a model has emerged that likens each chromosome to a gene string with the genes arranged in linear array. Each gene possesses a function distinct from that of the other genes. The alleles, or alternative forms of a *given* gene, represent variants of the particular function controlled by a gene. In time, only one of the family of alleles of a given gene is on a chromosome; in space, it is always in the same position on the same chromosome. This statement, as will be exposed in detail later, simply reflects the experimental truth that we can "map" a gene, locate its position on a particular chromosome, and know we can find it there when we wish in the progeny cells of the organism after many generations. Note that the number of different strings corresponds to the haploid number of chromosomes characteristic of a given species.

Now, the chemical substance of the gene is DNA. How is the DNA arranged on the chromosome? Moreover, the chromosome contains proteins, RNA, and a swarm of satellite molecules so that when the DNA gives orders it can have them fulfilled. How do we fit these in?

Before we go off on a speculative jag, things will make more sense if we first consider what it is that a gene does.

NEUROSPORA LIFE CYCLE

Let us consider in some detail inheritance in the fungus *Neurospora crassa*. This organism was chosen not merely because it has made welcome contributions to the authors' present high standard of living, but because its life cycle offers unique advantages for the study of gene action. (Fig. 3-1) (1) It is *haploid;* this means that the *expression* of a gene in neurospora is not complicated by the presence of an additional allele on a sister chromosome as in a diploid organism. (2) Under defined growth conditions, neurospora produces large numbers of asexual spores called conidia. These arise by mitotic divisions of haploid nuclei, and are available in large masses for experimental inspiration and manipulation. (3) Neurospora also has a well-defined and controllable *sexual* phase. Sexual reproduction occurs through fusion of two haploid nuclei to give rise to a diploid nucleus. The diploid nucleus then undergoes a reduction division by meiosis to give characteristic sexual spores (ascospores). Each ascospore can be isolated, induced to germinate and give rise to a plant, and thereby reveal the products of the mating. (4) Growth is fast. Many experiments can be done in a short time to appease the appetite of the researcher.

Neurospora normally grows as a spreading mycelium. You may already have seen it as a common contaminant on bread. The mycelium is composed of long individual branched strands called hyphae. In many fungi, the hyphae are strings of attached cells, but in neurospora the individual cells are not separated by complete cell walls. The cytoplasm is continuous throughout the hyphae and there are many nuclei in this common cytoplasm. All the nuclei are haploid.

Fig. 3-1. Life cycle of neurospora. The asexual phase in the life cycle occurs as a spreading mycelium called hyphae. Asexual reproduction can occur indefinitely through the formation of asexual spores, conidiospores. The sexual phase is initiated by fusion of strains of opposite mating type, indicated in the figure as mating types A and a. The sexual stage is carried out in a characteristic body, called a fruiting body. One nucleus of strain A fuses with a nucleus of strain a in the fruiting body and this diploid nucleus then undergoes meiosis giving rise to the characteristic sac-like structure called an ascus which contains the products of meiosis, ascospores. Ascospores, upon germination, give rise to the normal mycelial growth, and mating type shows segregation typical of a single gene difference.

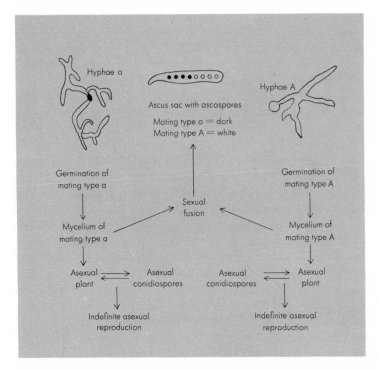

Vegetative or asexual reproduction occurs in two ways, by growth of the hyphae and repeated nuclear division, or by the formation of asexual spores called conidia.

In the sexual phase of neurospora, fusion of the two haploid nuclei occurs through the union of two strains of opposite mating types. These strains are alike in form, but they differ in mating type, as is shown by the fact that if strains of opposite mating type are grown together, characteristic sexual ascospores are formed. Strains of similar mating type, if grown together, do not form such spores.

MEIOSIS

The consequence of meiosis is a reduction in chromosome number from 2n to n. This is mandatory for sexual reproduction; otherwise, chromosomes would pile up, each generation doubling the number ad infinitum until life would be just a bowl of chromosomes. *When* meiosis occurs, however, is another matter. In man, a diploid organism, reduction to haploidy occurs only in the gametes prior to fertilization. Since neurospora lives out its allotted time as a haploid organism, meiosis occurs immediately after fertilization. It is important to re-emphasize that the chromosomes in a diploid cell exist in pairs, one member of each pair having been contributed by each parent. Human diploid cells contain 23 pairs of chromosomes, not 46 unlike chromosomes; the short-lived zygote of *Neurospora crassa* contains 7 pairs of chromosomes since each mating type has contributed one complete haploid set of 7.

As with mitosis, a copy of each chromosome is synthesized before meiosis begins. While each newly made chromosome remains attached to its model, the cell is now effectively 4n. The subsequent steps must ensure the reduction of one 4n cell to four n cells. Consider any given pair of chromosomes. Initially, and quite distinctly from what goes on in mitosis, the duplicated pairs of chromosomes show strong mutual attraction and come to lie very close to one another, probably within atomic distance, with the genes on one pair of chromosomes lining up next to their alleles on the other duplicated chromosome. (This arrangement permits exchange of chromosomal segments between partners—the genetic consequences of exchange will be discussed later.) Look at Fig. 3-2. What happens now is that the four chromosome strands (a tetrad) undergo two reduction divisions. At the first reduction division into two cells, the duplicated partners separate (segregate). At the second reduction division into four cells, the sister chromatids segregate. Each of the four chromosomes in each tetrad is now safely dissociated into one of four cells. Since each tetrad distributes itself independently of any other tetrad, haploid nuclei are produced that contain a full complement of chromosomes, but some are from one parent and some from the other.

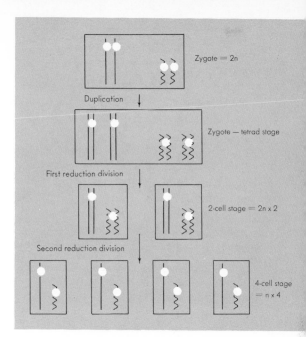

Fig. 3-2. The stages of meiosis in a diploid cell having two chromosomes. The chromosomes duplicate to give a 4n tetrad stage. Two reduction divisions then yield four haploid (n) cells.

Let us now consider the transmission of a specific trait in this organism. Neurospora normally grows as a spreading mycelium. A variant form is known which shows a button-like or colonial type of growth. The colonial type can be mated with a normal strain and their progeny examined. Upon dissection and germination of the eight ascospores resulting from such a cross, four will show the colonial growth of the variant and four the normal spreading growth. Half of the progeny are like one parent, the other half like the second. This would be expected if the trait—spreading growth versus colonial growth—is determined by a single gene. The normal strain has a gene (+) on one chromosome that determines spreading growth. The colonial strain has an alternative form of this same gene (c) that results in colonial growth.

As shown in Fig. 3-3, the fusion nucleus formed by the mating of these two strains contains two homologous chromosomes, one bearing the normal gene (+) and the second the colonial allele (c). At meiosis, these two homologous chromosomes pair and form a tetrad. At the first division, the replicated members of the pair are drawn to opposite ends of the ascus, and so segregate the (+) and (c) alleles from each other. The second division's spindles are then formed and segregate the sister strands, giving two nuclei having the (+) gene and two nuclei having the (c) gene. Each nucleus now undergoes a mitotic division to give four nuclei having the (+) gene and four nuclei having the (c) gene. This example clearly illustrates the important general principle that any trait in a haploid organism, such as neurospora, which is controlled by a *single gene* must show a one-to-one segregation in the progeny. This inevitably follows from the fact that there is equal genetic contribution by each parent in the formation of the zygote, and that each homologous chromosome shows equal replication.

RANDOM ASSORTMENT

The haploid nucleus of neurospora contains seven chromosomes. Specific genes are known to be carried on each of the seven chromosomes. This poses the question of how two genes, each carried on a separate chromo-

Fig. 3-3. The inheritance of a single-gene difference. Gene (c) determines a colonial-type growth, while its normal allele determines a spreading growth. The sequence at the left indicates the nuclear behavior. Note that segregation occurs at the first division, giving a 4:4 spore arrangement in the ascus. (After Beadle.)

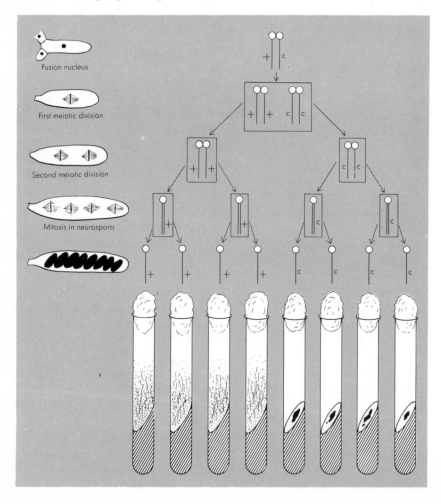

some, segregate during meiosis. We can answer the question by studying the transmission of two sets of traits, the traits being determined by genes on different chromosomes. For instance, assume that two alleles known to control spreading ($+$) versus colonial (c) growth are situated on chromosome No. 1. In addition, assume that a gene on chromosome No. 2 determines the formation of orange conidia ($+$), while an allele of this gene determines the formation of white, or albino, conidia (alb). Let us analyze the progeny that arise by crossing an orange colonial strain ($+$, c) by an albino spreading strain (alb, $+$).

First, some shorthand. The symbol $c/+$ stands for two alleles of one gene; each allele is on the chromosome contributed by one of the parents of the cross; $+$ is the general symbol for the "wild-type" allele; historically this refers to the allele found in nature, although modern usage has muddied history. The lower case letter, here (c), stands for the other allele, a mutant allele. The symbol (c, $+$) refers to two genes and signifies that one of the mating types has both the allele c of one gene and the wild-type allele, $+$, of the second.

We now return to the analysis. A large number of ascospores are isolated from the mating between ($+$, c) and (alb, $+$) and plants from each spore are grown and scored for the traits involved. First, we determine the distribution of each character *separately*. We find, as before, that about one half of the total plants are spreading, the other half colonial. As for color, analyzing the whole population again, one half are orange, the other half albino. Thus, each trait by itself segregates as a single gene difference. Now let us look at the combination of characters. Postulate: If the two genes are inherited independently of one another, then any portion of the population of ascospores we choose at random should give half albino and half orange plants. In particular, suppose we select only the colonial plants; one-half should be orange, one-half albino. Or, if we select the spreading plant, again one-half should be orange, one-half albino. The consequence is then obvious. If half of all the plants are colonial, and half of *these* are albino, then one-half $\times$ one-half, or one-fourth, of all the plants should be both albino and colonial (alb, c); similarly, one-fourth should be orange, spreading ($+$, $+$); one-fourth ($+$, c); one-fourth (alb, $+$).

Let us consider the independent inheritance of traits in terms of the mechanics of chromosome transmission.

The diploid cell, or zygote, formed by mating a colonial and an albino strain will contain 14 chromosomes consisting of 7 pairs. Chromosome-1 contributed by the colonial strain will carry the gene (c). Chromosome-1 contributed by the albino strain will carry the normal, spreading allele ($+$). Of the four cells containing the meiotic products of gene 1, half must contain the ($+$) allele, the other half the colonial (c) allele (see Fig. 3-3).

Chromosome-2 is distributed independently of chromosome-1. There-fore, half of the cells with the spreading (+) allele on chromosome-1 will get the orange (+) allele on chromosome-2, and half will get the albino (*alb*) allele on chromosome-2. This leads to the meiotic products (+, +) and (+, *alb*). The same analysis holds for the cells that contain the colonial (*c*) gene. Half of these will be (*c*, +), the other half (*c*, *alb*). Thus, the correspondence between trait inheritance and chromosome assortment stands up.

Note one other important fact. The parental strains in this cross were (+, *c*) and (*alb*, +). In the progeny, both parental strains (relative to these two genes) were recovered. In addition, and extremely important, two other classes were recovered (+, +) and (*alb*, *c*). These are recom-binant classes and point up one of the extremely important consequences of sex, namely, the redistribution of genes into new combinations.

LINKED GENES

Since many genes are carried on a single chromosome, we must also con-sider the transmission of two genes both of which are carried on the same chromosome. Such genes show the phenomenon of linkage, i.e., they are transmitted together more frequently than would be predicted by chance. Such genes do, however, undergo recombination as a consequence of cross-ing over. Consider two gene pairs, $p/+$ and $+/s$, both of which are carried on the same chromosome, with s farther from the centromere than p:

centromere p s

The centromere is the point on the chromosome where the sister strands remain attached prior to separation, and it is also the point where the fibers from the spindle attach and draw the strands to opposite poles.

If the strains (*p*, +) and (+, *s*) are mated, and progeny ascospores isolated, the majority of the spores upon germination would be of one or the other of the two parental types, either (*p*, +) or (+, *s*). However, recombinant strains are also found, i.e., strains that are normal with respect to both traits (+, +) as well as the double mutant carrying both the *s* and *p* traits (*p*, *s*). The two *recombinant* classes are invariably present in progeny in equal proportion. The total *number* of recombinant progeny, however, in this case, is less than that predicted on a basis of random or independent transmission. Remember, if these two genes were transmitted independently of each other, 25 per cent of the progeny would be like one parent (*p*, +), 25 per cent like the second (+, *s*), 25 per cent would be normal (+, +), a recombinant class, and 25 per cent would be the double mutant (*p*, *s*), the second recombinant class. Therefore, of the total prog-eny 50 per cent would be parental and 50 per cent recombinants.

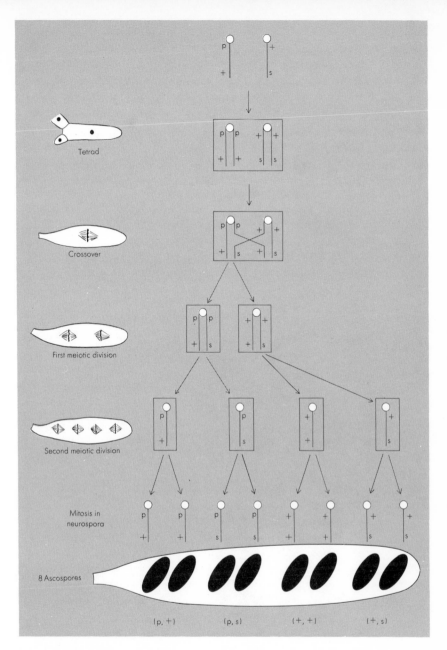

Fig. 3-4. Exchange between chromatids. Note that segregation occurs after the second division, resulting in a 2:2:2:2 spore arrangement in the ascus.

The hallmark of linkage is that the total number in the two recombinant classes is significantly less than 50 per cent of the progeny. As shown in Fig. 3-4, recombination of linked genes results from an exchange of segments between nonsister strands in the closely paired tetrad formed in the zygote by homologous chromosomes. The exchange, termed crossing over, occurs before the first reduction division. As can be seen in the figure, if a break occurs between genes *p* and *s,* exchange of the terminal segments will result in one strand carrying the two normal alleles (+, +). The other homologous strand will carry the two mutant alleles (*s, p*). Note carefully from the figure how the two meiotic divisions then segregate the four different genotypes.

Fig. 3-5. A schematic diagram of the genetic events concerned with a first-division segregation and with a second-division segregation.

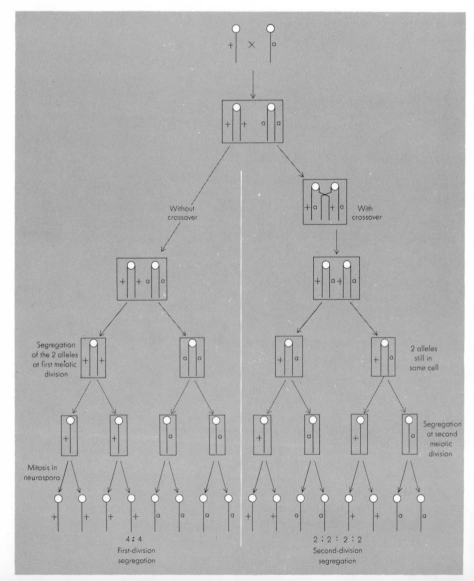

Whether or not a crossover between linked genes has occurred can be readily determined in neurospora by noting the spore arrangement in the ascus. If no crossover occurs between two genes, the first meiotic division will result in their separation, the ascus will show a 4:4 arrangement, and the two classes will be identical to the parents. If, however, a crossover has occurred, segregation will not occur until the second division, it will result in a 2:2:2:2 segregation (arranged either as 2:2:2:2 or 2:4:2), and the two parental and two recombinant classes will be present. This is diagrammed in Fig. 3-5. In the photograph (Fig. 3-6, top), one can clearly see both crossover and noncrossover asci.

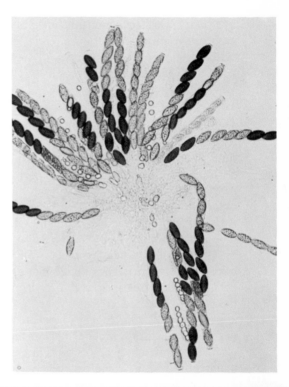

Fig. 3-6. (Top) Asci produced by crossing a normal strain of neurospora with a strain that shows delayed maturation. Note both first- and second-division segregations. (Courtesy Dr. David R. Stadler.) (Bottom) A schematic representation of the gene map of a chromosome of Neurospora crassa. **The numbers after a gene indicate different genes affecting the formation of the same end product. The relative distances on the map are derived from recombination data as discussed in the text.**

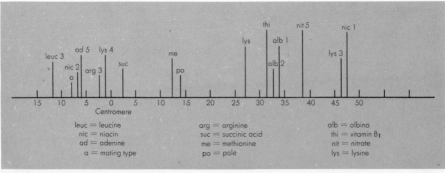

28

GENES AND BIOCHEMICAL REACTIONS

The farther away a gene is from the centromere, then, the greater the chance of a crossover between the gene and centromere, and the greater the chance of a 2:2:2:2 (second-division) segregation. The frequency of first- and second-division segregation thus permits the ordering (mapping) of genes relative to the centromere, for the gene closest to the centromere will have the largest ratio of 1st/2nd division segregation, while the gene farthest away will have the smallest. The fact that the chance of a crossover in any given part of a chromosome is about the same for any other part of the chromosome means that the farther apart two genes are, the greater the probability that a crossover can occur. The greater the probability that a crossover will occur, the higher the frequency of recombination. Thus the frequency of recombination is a measure of the relative distance between two genes. This fact permits the preparation of genetic maps. A map of one chromosome of neurospora is given in Fig. 3-6 (bottom). For a discussion of the details of the construction of such a map, consult Chapter 9 of the text by Srb and Owen.*

MUTATION

Gene configuration is not fixed. As has been mentioned, a given gene can exist in several different forms (alleles). A change of a gene from one form to another is called mutation. Mutation is of paramount importance to the study of heredity, for it must be re-emphasized that we can be aware of the existence of a gene only when more than one allele of that gene is observed. The rare mutation that occurs spontaneously in nature provides variation and permits the evolution of new and novel forms. Mutation frequency, however, can be increased experimentally. This was first shown by H. J. Muller, using X-rays. X-rays and other ionizing radiations increase mutation frequency, as do many other agents, e.g., heat, ultraviolet light, and chemicals such as nitrous acid and nitrogen mustard (similar to the mustard gas used in World War I). These and other mutagenic agents are extremely useful in genetic research, since they permit us to explore the possibilities for variation that are inherent in genetic material and to utilize many mutant genes as tools in experiments. The molecular basis of mutation will be discussed extensively in Chapter Five.

GENES AND NUTRITION

We are now in a position to consider a problem of primary contemporary interest and research, the mechanism of gene action. The genetic traits we have discussed up to now have been mainly morphological, but we know that a biochemical basis must underlie all such phenomena. Cells of a colonial form must differ from cells of a spreading form in biochemical characteristics, even though the nature of such differences is still unknown.

* A.M. Srb and R.D. Owen, *General Genetics* (San Francisco: Freeman, 1952).

Such a statement implies that gene action ultimately must be expressible in terms of cell chemistry. To get at this problem, we obviously need traits that can be studied biochemically. So-called nutritional traits satisfy this requirement. These were first intensively studied in microorganisms, and here we got the first clear answers. To grasp how these experiments were used for the analysis of gene action, however, requires some understanding of cellular biochemistry.

The basic characteristics of cellular biochemistry pertinent to a discussion of genetics will be briefly reviewed here. For a more detailed description, the reader is referred to the book by W. D. McElroy in this same series.*

Comparative biochemistry has taught us that there are great similarities between cells of phylogenetically different organisms. A surprising biochemical unity is, in fact, found throughout the living world. All living cells are similar in composition and contain the same classes of chemical compounds. Despite variation, all cells are built from the same building blocks, and all living things need essentially the same basic nutrients for growth. They require a source of carbon, a source of nitrogen, an energy source, and a source of minerals. From these nutrients, cells synthesize their major chemical components—such as proteins, carbohydrates, DNA, RNA, and lipids—as well as the chemical compounds—such as amino acids and purine and pyrimidine bases—which compose these macromolecules.

A cell synthesizes its vital components by a series of well-defined chemical reactions. Most of these reactions require energy. A cell, therefore, carries out another series of reactions in which energy can be trapped and forced into synthetic reactions. A constant interplay thus exists between energy-releasing and energy-requiring reactions. The astonishing fact that the cell can carry out all its necessary reactions under conditions of constant temperature and pressure is due to the presence of specific catalysts, enzymes. (Remember, in the laboratory the chemist can use temperatures and pressures over many thousandfold ranges, while the human cell must do its biochemistry at 37°C and a pressure of about 15 lb/in.²) Enzymes, like other catalysts, can markedly increase or decrease the rates of reactions without themselves undergoing change. That is why only very small amounts are needed per cell.

All enzymes are proteins, and many of them require a cofactor (some are vitamins), usually a small molecule or a metal ion, to carry out their catalytic role. Enzymes are also characterized by specificity, i.e., a single enzyme can catalyze only a single reaction or class of reactions. In Fig. 3-7, a series of biosynthetic reactions illustrates these points. The amino acid arginine is synthesized from citrulline, citrulline in turn from ornithine, and ornithine from glutamic acid. Glutamic acid can be made from glucose by a long series of reactions. Thus arginine is formed from glucose by a large

* W.D. McElroy, *Cell Physiology and Biochemistry*, 2nd ed. (Englewood Cliffs, N. J.: Prentice-Hall, Inc., 1964).

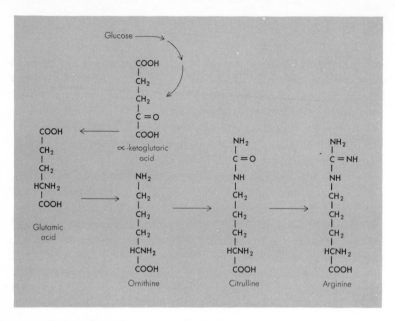

Fig. 3-7. The biochemical reactions required for the formation of arginine from glucose. The reactions in the conversion of ornithine to arginine are unique to arginine formation, while the earlier steps are common to the formation of other substances.

number of chemical reactions, and each of these reactions requires the participation of a specific enzyme. A similar series of events is found in the synthesis of the amino acid tryptophan, although different intermediate compounds are involved.

One last point should be noted. Although a majority of the reactions involved in the formation of arginine are common to the formation of many compounds, there are a few terminal steps which are unique. (These will be pointed out shortly.) This is generally true of the biosynthetic sequences involved in the formation of most end products.

Living cells vary considerably in their synthetic powers, depending on which enzymes they possess. Obviously, if a cell cannot synthesize a certain essential compound, it requires that compound as a nutrient. Therefore, dietary requirements reflect enzymatic capabilities. Some organisms can synthesize all their amino acids, vitamins, and nucleotides. They get the energy to do this from relatively simple compounds like sugar. Many microorganisms are of this type. Plants are capable of synthesizing their amino acids, etc., from even simpler compounds, but they require sunlight as their energy source. Other organisms lack the necessary enzymes to make many of their essential constituents, and so require them in their diet. Man, for instance, cannot synthesize his own vitamins, and can synthesize only about one-third of his amino acids.

Study of nutritional differences points to an important fact for the study of genetics. Essential compounds, like amino acids and vitamins, are synthesized by a series of well-defined steps. An organism can lose the ability to synthesize such compounds and the loss need not prove fatal, since these substances can be provided in the diet. As a test, then, of whether or not genes control cellular reactions, reactions involving the synthesis of such substances prove to be excellent means of getting at the basis of the genetic control of cellular biochemistry. For example, using an organism which can synthesize all its needed amino acids, we can determine whether a given mutation results in a nutritional requirement for a given amino acid. If it does, does this nutritional requirement induced by mutation result from the loss of ability to carry out a specific biochemical reaction? Finally, is this loss inheritable?

Experiments of this sort were initiated by the geneticist George W. Beadle and the biochemist Edward L. Tatum, for which they were awarded the Nobel Prize in medicine in 1959. Their experiments with the fungus *Neurospora crassa* gave conclusive proof that mutation can result in nutritional alteration, that the basis of the nutritional alteration rests on the loss of the organism's ability to carry out specific biochemical reactions, and that the loss was a heritable trait.

To see the basis of this conclusion, consider an actual experiment. Neurospora utilizes sucrose as both a carbon and an energy source. It can utilize nitrate as a nitrogen source and requires one vitamin—biotin. Neurospora, therefore, grows on a medium containing sucrose, nitrate, biotin, and minerals (minimal medium). Microconidia (the uninucleate haploid spores of neurospora) can be treated with a mutagen such as X-rays, and these irradiated spores can then be plated out on an enriched minimal medium, i.e., a minimal medium to which the known vitamins and amino acids have been added. This supplementation permits growth and survival of nutritional mutants. Under appropriate conditions, these spores will grow into discrete colonies. As a test of whether or not a mutation which affects a nutritional trait has occurred, a portion of each colony can be put into minimal medium. Under these conditions, some of the colonies are found to be no longer capable of growing on a minimal medium, although they are still capable of growing on a supplemented medium. The strains are then tested further to determine which of the supplements in the enriched medium is required, and many strains are found to require only one substance in addition to the needed minimal medium. For instance, strains are found which require arginine, others tryptophan, and still others the vitamin niacin. Treatment with a mutagenic agent can therefore lead to the formation of strains having altered nutritional characteristics.

Are such nutritional differences heritable differences or are they due to a nongenetic or nonchromosomal change? This question can be readily answered by means of the inheritance test. If an arginine-requiring strain

$(A-)$ is crossed with a parental strain $(A+)$, the eight ascospores of a single ascus, representing the four products of meiosis, can be isolated and grown. When these spores are germinated on a medium containing arginine, all eight spores grow. If these eight cultures are now transferred to a medium lacking arginine, four of the strains will grow and four will not. Half of the spores are arginine-dependent. This trait, therefore, segregates as predicted for a single-gene difference [refer back to Fig. 3-3 and in place of (c) substitute $(A-)$]. This type of inheritance is characteristic of all nutritional mutants. Alteration of a single gene can result in loss of ability to form a specific essential compound.

GENES AND BIOCHEMICAL REACTIONS

To pinpoint the biochemical nature of this induced nutritional alteration, we must look at the cellular reactions involved in the formation of the required end product. Let us continue with arginine. The biosynthesis of arginine requires the formation of glutamic acid from glucose. Glutamic acid in turn is converted to arginine through the formation of ornithine and citrulline, reactions unique to the formation of arginine. (See Fig. 3-7.) The synthesis of arginine actually involves the formation of compounds closely related to ornithine and citrulline, but, for the sake of clarity, we will discuss its formation as described above. Mutant strains which *require* arginine for growth can be tested for their ability to grow on these compounds, compounds uniquely required for arginine synthesis. The question, then, is will arginine-requiring mutants grow on ornithine and citrulline as well? (See Fig. 3-8.) We find that some of the mutants will grow only when supplied with arginine. Others will grow on citrulline, and still others on citrulline or ornithine.

These mutants are not biochemically identical. One class appears to be unable to form arginine because of its inability to carry out the conversion of citrulline to arginine. A second class of mutants appears unable to form arginine due to an inability to carry out the conversion of ornithine to citrulline. They *can,* however, form arginine from citrulline, since they can use citrulline for growth. The third class lacks the ability to form arginine, through inability to carry out the conversion of glutamic acid to ornithine. Each of these mutants seems unable to carry out one specific biochemical reaction. This suggests that mutation of a *single gene* leads to the loss of the ability to carry out but one biochemical reaction.

If these mutants are examined genetically, we find that all the strains that cannot convert *citrulline to arginine* are genetically similar. All these strains arise by mutation of the same gene. Similarly, all the strains that cannot convert ornithine to citrulline involve mutation of one gene, but a gene distinct from the one just mentioned. Still a third gene is found in which mutation results in the loss of ability to convert glutamic acid to ornithine.

There appears to be a clear association between a specific gene and the ability of the organism to carry out a specific biochemical reaction. This has just been shown for arginine.

Similar results are found for mutations that affect the formation of tryptophan, as well as for mutations that affect the formation of other vital substances. Not only is there an association between one gene and one biochemical reaction, but there is also specificity: A specific gene controls a specific biochemical reaction. Since each of these reactions involves enzyme catalysis, the one–one correlation actually reflects a relationship between genes and enzymes, rather than between genes and biochemical reactions. This correlation established by study of nutritional mutants is known as the one-gene, one-enzyme hypothesis and states that one gene controls the formation of one enzyme. The answer to one question thus shapes another. How can a gene control the formation of a specific enzyme? To answer this, we must consider in greater detail the relationship between genes and proteins.

Fig. 3-8. The genetic control of the biochemical reaction concerned with arginine formation. Compounds which can be used for growth by the various genetic classes are indicated on the left side.

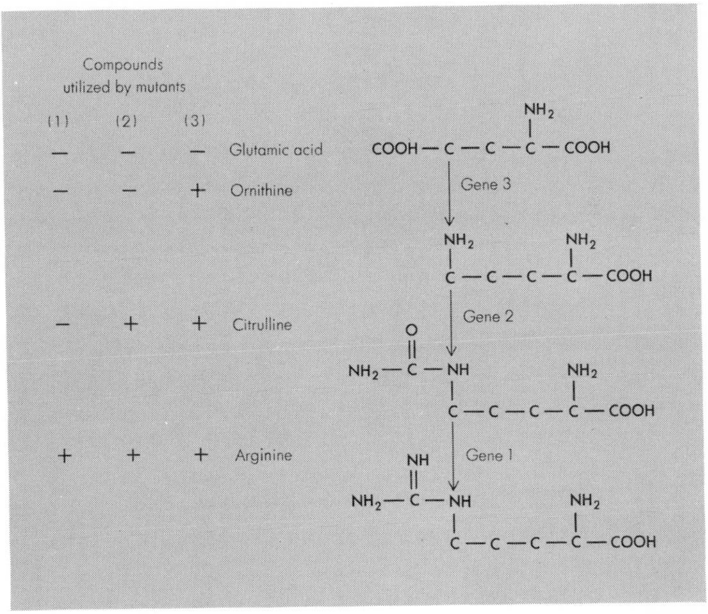

Genes and Enzymes

CHAPTER FOUR

In the previous chapter, it was stated that specific genes control the formation of specific enzymes. This was based largely on the observation that mutation of a specific gene may result in the loss of ability to carry out a specific reaction. Additional evidence may be obtained by testing for enzyme activities in the test tube. Cell-free extracts can be prepared both from a mutant strain and the parental strain from which the mutant was derived, and the extracts examined for the presence of the enzyme. In general, analysis of the extracts confirms that a mutation which results in the loss of ability to carry out a specific reaction also results in loss of the enzyme *activity* that is required for the catalysis of the reaction.

This clear relationship between gene and enzyme suggests that an understanding of the nature of this relationship might well go a long way toward explaining how genes act. Let us probe further. It would be of interest to know, for example, whether the result of mutation is simply a failure to form enzyme or is a consequence of a change in the structure of the formed enzyme. Does a gene exert control by regulating the rate and other quantitative characteristics of enzyme formation, or does it control the architecture of the enzyme, i.e., the sequence of amino acids or their relative proportions? For reasons of experimental expediency, these questions have been studied in detail only in microorganisms, and our discussion must deal with these organisms. However, it should be borne in mind that the principles deduced from the study of microorganisms are general and pertain to all living things, including man, as will be documented in a subsequent chapter.

TRYTOPHAN SYNTHETASE

A number of different genes and enzymes have been closely analyzed during the past ten years. Since all the systems suggest a similar relationship of gene to enzyme, we can concentrate on one well-described system. The enzyme tryptophan synthetase has been studied in detail in fungi and bacteria. This enzyme is required by these organisms for the synthesis of the amino acid tryptophan. The formation of tryptophan involves the synthesis of a unique intermediate, indoleglycerol phosphate (InGP). The intermediate reacts with serine as shown in Fig. 4-1. A substitution occurs between serine and the triose phosphate portion of InGP to give tryptophan and triose phosphate. This, the terminal step in tryptophan formation, is catalyzed by tryptophan synthetase (t'ase).

Mutations affecting the formation of t'ase can be recognized by two criteria: (1) They are mutations which result in a nutritional requirement for tryptophan, and (2) for growth, tryptophan cannot be replaced by InGP or the precursor of InGP, anthranilic acid. With neurospora, a large number of conidiospores can be exposed to a mutagenic agent, e.g., nitrous acid, and we can select out many independently arising mutant strains of the above type. None of these mutant strains has measurable t'ase activity in cell-free extracts. Such an allelic series can be used in a number of ways to examine the nature of the gene-enzyme relationship.

One such problem is: In how many regions of the entire genetic structure (the genome) of neurospora can mutation occur and cause loss of

Fig. 4-1. The terminal step in tryptophan biosynthesis, catalyzed by tryptophan synthetase.

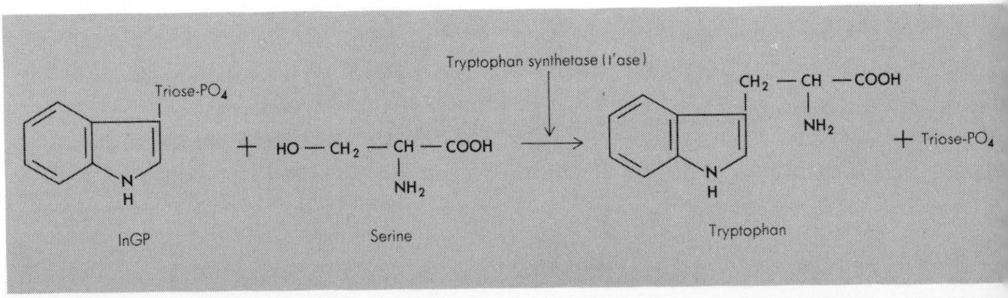

t'ase activity? Is there a single region—or are there many regions? This problem can be answered by crossing all the mutant strains with one other and determining whether tryptophan-independent progeny arise. If there are two areas on separate chromosomes, or two distinct areas on one chromosome, both of which control t'ase formation, frequent tryptophan-independent progeny will be observed as a result of recombination when the strains are crossed. If, on the other hand, the areas are identical or very closely linked, few if any tryptophan-independent progeny should be found. Crosses between members of the mutant series clearly established that they are all mutations of the same area of the genome. In neurospora, this area is on the second linkage group.

It can be stated categorically, then, that there is a single region in the genome which directs t'ase formation, and in which mutation can occur to cause loss of ability to form functional enzyme.

The next problem is: Does mutation in this genetic area decrease the rate of formation of the enzyme, or does mutation result in structural changes in the formed protein, or both? This problem can be experimentally examined by determining whether, as a consequence of mutation, no enzyme is formed, whether enzyme is slowly formed, or whether the altered gene still directs the formation of a specific protein but one which lacks the catalytic activity of t'ase. To obtain evidence on the last point, we must be able to recognize and pinpoint a given enzyme by properties other than its catalytic activity. This can be done by using immunochemical techniques.

DETECTING CRM

It has been known for a great many years that animals can form proteins, called antibodies, in response to a challenge by certain foreign materials (antigens). For example, many of you have been immunized against diphtheria and poliomyelitis. You react by synthesizing antibodies which collect in the serum portion of your blood. If you then come into contact with either diphtheria toxin or the polio virus, an antigen-antibody reaction takes place. The toxin is neutralized and rendered harmless, while the virus is prevented from infecting your nerve cells. Antibody formation can be elicited by many substances, including proteins. Important to our discussion is the fact that antigen-antibody reactions are highly specific. (Note how specificity is basic to so many biological systems, enzymes, antibodies, viruses.) Antibodies formed against one protein will react only with that protein, or proteins which are *structurally related*. It is the specificity of the antigen-antibody response that makes this a remarkably useful tool in studying the consequences of mutation at the enzymatic level.

Let us extend just a little the idea of structural relation as it pertains to the antibody-antigen reaction. An antibody made in response to a protein antigen is not made to the whole protein antigen; the antibody is made to and reacts with just a small area of the antigenic protein and each antigen generally contains a number of antigenic areas or sites. An antibody reacts only with a particular antigenic site. Structural relation, then, means that if the same spatial arrangement of atoms exists on another protein, the antibody will also react with it. When a protein is injected into an animal, therefore, antibodies are formed that react with different parts of the protein. If the same antibodies react with another protein, then it can be inferred that some portions of the two proteins are similar in atomic arrangement.

T'ase is, of course, a protein and it can be purified. If a preparation of pure t'ase is injected into rabbits, specific antibodies are formed in response to a challenge by this antigen. These antibodies are present in the rabbit serum and can be recognized by the fact that serum taken from the immunized rabbit completely inhibits (neutralizes) catalytic activity when added to the enzyme in the test tube. Serum taken from the same rabbit before immunization has no effect on enzyme activity (Fig. 4-2). The antibodies are specific for t'ase in that they are without effect on other enzymes. (It can be shown that not all the antibodies made against t'ase are neutralizing, but this is consistent with our knowledge that the catalytic site is only a small portion of the enzyme.)

We now have two ways to identify t'ase: (1) by its ability to catalyze InGP + serine $\longrightarrow$ tryptophan, and (2) by its ability to react with t'ase antibodies. If these two characteristics represent two different properties of the protein, we have a tool that permits the recognition of the t'ase protein, even in the absence of catalytic activity.

The point of all this is that neutralizing t'ase antibodies can be used to detect the presence of a protein which is structurally similar to t'ase, but

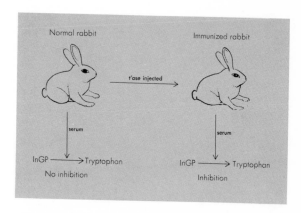

Fig. 4-2. Enzyme neutralization by antibodies.

lacks the catalytic properties of t'ase. We will designate such a protein as CRM. To repeat, we want to know whether the mutant t'ase gene can no longer direct the formation of a protein, or whether it still directs the synthesis of a related but enzymatically inactive protein. This can be examined as follows (see Fig. 4-3). A cell-free extract of a mutant strain can be prepared. A known amount of t'ase antibody is added to the extract. If a protein is present in the extract which can react with these antibodies (CRM+), it will do so and tie them up. If some t'ase is added, there will be no antibodies to inhibit the enzyme, which can then go ahead and make tryptophan from InGP, serine, and pyridoxal phosphate. Suppose the cell-free extract does not contain a protein which reacts with the neutralizing antibodies. The antibodies will, of course, remain free. They will neutralize added t'ase and no tryptophan will be formed from InGP, etc. Ultimately, therefore, as indicated in Fig. 4-3, the formation of tryptophan shows the presence of protein in the original cell-free extract which can react with and remove t'ase antibodies. No tryptophan formation shows the lack of such a protein in the extract.

T'ASE CRM

If t'ase-less mutants are examined in this way for the presence of a protein which is serologically related to the parental enzyme in terms of neutralizing antibody, it is found that mutation may give rise to strains unable to form such a protein. Such mutants are called CRM−. However, this is not true of all the mutants. The majority of the mutants do form a protein, and in normal amounts (CRM+). This protein will react with neutralizing t'ase

Fig. 4-3. The test for the presence of a protein which is related to the t'ase of the parental strain.

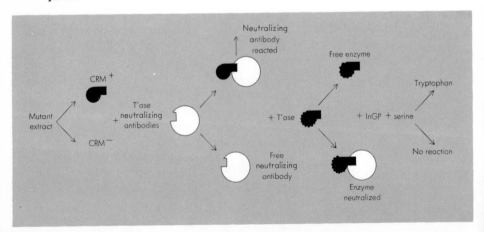

antibodies but is unable to catalyze the conversion of InGP to tryptophan. Thus mutation may result in loss of catalytic activity without resulting in loss of ability to form a structurally similar protein. This suggests that mutation, in this instance, is not simply controlling the rate of enzyme formation. Rather, it is affecting enzyme structure.

This fact can be seen even more convincingly in the following way. T'ase catalyzes the reaction InGP and serine ⟶ tryptophan (1). However, this same enzyme can catalyze two related reactions, indole and serine ⟶ tryptophan (2), and InGP ⟶ indole (3). The three reactions catalyzed by t'ase are shown in Fig. 4-4. Reactions 1 and 2 both require serine and pyridoxal phosphate, while reaction 3 has no cofactor requirements. If mutation alters enzyme structure, we might predict that mutations of this gene should be found which will result in loss of catalytic activity in one or some combination of these reactions, but not in others, e.g., loss of reactions 1 and 2 but not of 3, or loss of reactions 1 and 3 but not of 2. If the gene, however, is controlling enzyme formation by controlling the rate of formation, we would predict that mutation in every case should decrease all three reactions equally.

Mutations which result in loss of reactions 1 and 2, but not of 3, are found and easily detected, since such strains characteristically require tryptophan for growth and accumulate indole in their culture medium. All such mutants are CRM+ (i.e., react with neutralizing t'ase antibodies). Their CRM, on isolation and purification, can catalyze reaction 3 in the test tube.

Similarly, mutations which result in loss of reactions 1 and 3, but not of 2, are also found and again are easily detected, since such strains charac-

Fig. 4-4. The reactions catalyzed by t'ase.

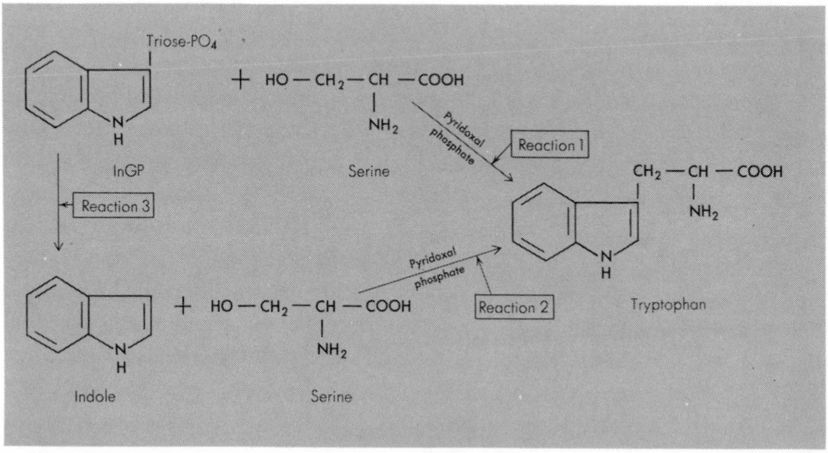

teristically require tryptophan for growth, but they can grow equally well if given indole instead. Again, all such mutants are CRM+ and their CRM, on isolation and purification, will catalyze reaction 2 *in vitro,* i.e., outside the cell.

Mutations which result in loss of reactions 2 and 3, but not of 1, have not yet been observed, since reaction 1 is the reaction normally involved in tryptophan formation. Mutations of this class would not require tryptophan for growth and would pass undetected. To pick up a mutant of this kind, we are faced with the alarming prospect of making extracts of thousands of normal strains to find one incapable of carrying out 2 and 3. Similarly, mutations which result in loss of reaction 1, but not of 2 and 3, are also difficult to detect.

From such experimental observations, we can conclude that mutations in the *genetic region controlling t'ase formation* result in structural alterations of the formed protein, rather than in quantitative alterations such as rate of enzyme formation. This being true, we must conclude that the genetic area in some way controls the structure of the formed enzyme.

PROTEINS

Tryptophan synthetase, like all proteins, is basically a chain of amino acids. As previously stated, proteins are composed of 20 amino acids in varying amounts, the amounts and their distribution in the chain being characteristic for a given protein. The amino acids in a protein are linked by covalent peptide bonds, and the amino acid sequence is called the primary structure. The amino acid sequence of the enzyme, ribonuclease, which catalyzes the hydrolysis of RNA, is shown in Fig. 4-5.

The basic or primary structure of all enzymes is made up in a fashion similar to that of RNase. RNase is a relatively small molecule with a molecular weight = 13,683, while the molecular weight of t'ase is about 100,000. This difference, however, simply means that although RNase is composed of 124 amino acid residues, t'ase is composed of about 1,000. Detailed investigation of proteins suggests that the active catalytic sites of an enzyme consist of only a few amino acids; that is, the site on the enzyme surface where InGP combines consists of but a few of the 1,000 amino acid residues. The same is true for the serine site. For catalysis to occur, a very specific orientation of these two sites with respect to each other is required. To provide for this, an enzyme molecule does not exist as a linear extended structure. Rather, each enzyme has a precise folded structure. Experimental evidence at the present time suggests that the folding characteristics of a protein may be determined by the amino acid sequence. This conclusion may not withstand the test of time, but, for the sake of simplicity, we will assume that the amino acid sequence does determine the entire three-dimensional structure of an enzyme.

ENZYME ALTERATION

Since mutations appear to result in alteration of enzyme structure and since enzyme structure in turn is determined by its amino acid sequence, it follows that mutation must alter the amino acid sequence of the product enzyme. Can this be shown? The following discussion will enable us to answer this question.

$$H_2N-CH-C-OH \quad + \quad H_2N-CH-C-OH$$

with R, O above the first and R_1, O above the second

$$\downarrow$$

$$H_2N-CH-C-N-CH-C-OH \quad + \quad H_2O$$

with R, O and R_1, O above

Peptide bond

Fig. 4-5. (Right) The typical bond uniting the amino acid. (Bottom) The amino acid sequence of bovine pancreatic ribonuclease.

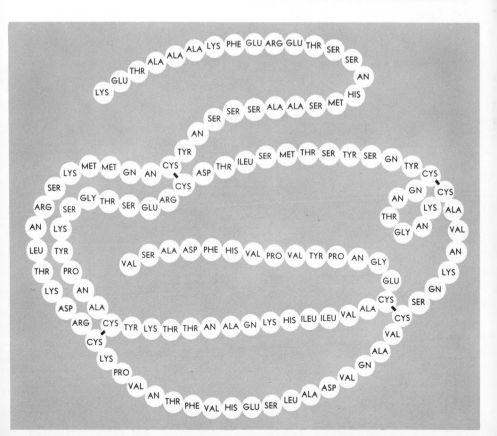

As mentioned earlier, a number of differences can be readily detected among mutants which are all alike in requiring tryptophan for growth and in lacking an enzyme catalytically active in converting InGP to tryptophan. Some of the mutants are CRM^-, others CRM^+. Some of the CRM's catalyze reaction 3, others reaction 2, still others neither. Of the CRM's which catalyze reaction 3, some require one cofactor, others two, still others none. If we examine the CRM's carefully, using quantitative immunochemical techniques, we find differences between them. If we compare them in terms of the temperature that is optimal for their activity, we find differences. In fact, after careful study, using many diverse criteria of comparison, we are left with the impression that few truly identical alleles are found. Each allele is functionally unique. This means that the genetic area controlling the formation of this enzyme is mutationally complex. It does not consist of a single site, alteration of which produces an identical set of mutants; rather, the region must consist of a large number of mutational sites, alteration of each one of which can produce a slightly different effect.

As we see in Fig. 4-6, we can mentally expand the genetic region and subdivide it into many mutational sites. This is what one would predict, if the gene controls the amino acid sequence and if each mutational site were to determine one amino acid of the sequence. Mutation could result in alteration of the site in such a way that were the site originally to direct the insertion of the amino acid valine into the sequence, it might now replace it

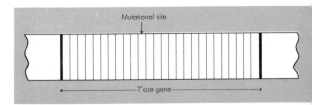

Fig. 4-6. Mutational sites within the t'ase gene.

with lysine. This change in sequence would be expected to have an effect on the folding characteristics of the molecule. There would be, as a consequence of this one change, a profound effect on the activity of the product formed. And we would expect a large number of such mutational sites, and almost endless variation in structure. This is what is found. To sum up, the genetic evidence suggests that the gene controls the amino acid sequence and that mutation results in alteration of sequence.

What do we find if we tackle the same problem by protein chemistry? Can we, by examining mutant protein, actually find amino acid substitution? This is not an easy task. The fact that an enzyme contains a total of several hundred amino acids of 20 different kinds makes the determination of amino acid sequence a particularly thorny problem. Fortunately, certain short cuts now permit "fingerprinting" of a protein. These methods enabled Vernon Ingram to demonstrate that there is a *single amino acid* difference between two genetically distinct, naturally occurring hemoglobins in man. The fingerprinting method can be readily shown in the case of RNase.

FINGERPRINTING

The enzyme trypsin catalyzes the cleavage by water (hydrolysis) of certain peptide bonds in proteins, namely, those in which one of the amino acids in the bond is either lysine or arginine. This action yields protein fragments called polypeptides. Treatment of RNase with trypsin will break the molecule into a number of polypeptides. These can then be separated from one another by a method called paper chromatography. The rate of migration on paper is characteristic of each fragment. If, for instance, two different trypsin digests of RNase of bovine origin are chromatographed and the two papers compared, we find similar patterns of superimposable polypeptide spots. If, however, a digest of bovine RNase is compared with that of horse RNase, the polypeptide spots are not superimposable, indicating that there are differences in the amino acid composition of the fragments and thus differences in the amino acid sequence of cow and horse RNase. Consider the sequence:

$$\text{val—ala—gly—lys—tryp—val—ser—arg—glu—ala—ser—gly}$$
$$\begin{array}{cccccccccccc} 1 & 2 & 3 & 4\uparrow & 5 & 6 & 7 & 8\uparrow & 9 & 10 & 11 & 12 \end{array}$$
$$\qquad\qquad\quad \text{Trypsin} \qquad\qquad\quad \text{Trypsin}$$

Treatment of this peptide with trypsin will break it into three fragments, and, on chromatography, we find three discrete peptide fragments. However, if the No. 9 glutamic acid is substituted with glycine, the No. 3 fragment will now occupy a different position on the paper as a consequence of its change in composition. If we compare the chromatographic profiles of these two, we find that the No. 3 peptide of the first preparation has disappeared, and we have in place a new fragment at a different spot—an elegant, yet simple method of showing amino acid differences.

A fingerprint comparison of mutant and normal t'ase has been carried out by Charles Yanofsky. Mutant CRM's from the bacterium *E. coli* were purified and subjected to fragmentation by trypsin. The peptide fragments were then separated from one another by appropriate chromatography. Chromatograms of the parental t'ase were prepared in exactly the same way, and the two fingerprints compared. Some mutant CRM's were found to have fingerprints identical with that of the parental enzyme. More important, however, many mutants were found whose fingerprints differ. In each instance, the difference was only one fragment, i.e., the absence of one parental fragment with the appearance of a new nonparental fragment.

The fragment involved was characteristic of the mutant, showing that the change in the amino acid sequence differs in mutants of independent origin.

That a peptide difference is found means that the mutant CRM differs from the parental enzyme in amino acid composition. In fact, the exact difference is known in most cases. For instance, the parental enzyme has a glycine that in one mutant has been replaced by arginine. Another mutant has replaced tyrosine by cysteine. Thus, mutation and amino acid replacement can be beautifully documented. But we also find one additional and important fact. As was indicated in Fig. 4-6, mutational sites are drawn as linearly arranged in the gene. Let us also take a look at the atomic arrangement of the uncoiled protein fragment.

Glycine   Aspartate   Alanine   Many amino acids   Tyrosine   Serine

Note that at one end there is a free amino group (NH_3^+) and at the other end the free carboxyl group ($-\overset{\displaystyle O}{\underset{\displaystyle \|}{C}}-O^-$). Assume the amino acid glycine, representing the amino or N-terminal end, is determined by the left-hand portion of the gene and the C-terminal amino acid, serine, is determined by the right-end of the gene. If a mutant found by genetic tests to map at the left end of the gene results in an amino acid replacement near glycine, while a mutant mapping at the right has an amino acid replacement near the C-terminal serine, we would conclude that the gene and its product enzyme, when stretched out, are co-linear. That's the way it seems to be. The evidence is a little sketchy as yet, but the authors' bias is that it is true, and that there is co-linearity between the DNA of the gene and its product enzyme.

A world of caution—finding CRM's which show no fingerprint difference is not necessarily proof of a lack of chemical difference. A chemical difference does exist, but is not detectable by present methods. Present methods will only detect alterations that create a difference in electrical charge or in chromatographic properties. Thus, substitution of glutamic acid (a dicarboxylic acid with a negative charge) by lysine (a diamino acid with a positive charge) can be detected, while substitution of leucine by valine (both uncharged) could not be detected.

We thus know that mutation at the tryptophan synthetase locus results in structural alteration of the formed enzyme because of differences in amino acid sequence. We know futher, from this and other loci, that alteration of a single mutational site results in the replacement of but one amino acid, and finally that there is co-linearity between the DNA of the gene and its product enzyme.

In summary, all the present experimental evidence clearly indicates that the genetic region controlling the formation of tryptophan synthetase controls the structural characteristics of this enzyme, i.e., the primary amino acid sequence. It is true for this enzyme, and study of other genes and other enzymes leads to a similar conclusion. One function of genetic material, therefore, involves the determination of enzyme structure. As will be discussed later, this conclusion does not force the further conclusion that all genetic material must act in the same way. Other genetic areas do play a role in the quantitative regulation of enzyme formation, but most enzymes probably have a gene that acts by controlling structure, and this action of genetic material should thus be explored further.

It is now apparent that a gene which controls the structure of an enzyme must be able in some way to serve as an amino acid code. In addition, it must be able to transmit this code to enzyme-forming centers. Since the chemical basis of heredity appears to reside in the compound DNA, we look to the properties and characteristics of DNA for hints to the solution of these new problems.

Genes in Action

CHAPTER FIVE

One of the interesting vices of scientists, not exempting biologists by any means, is their compulsion to shoehorn large masses of facts into small concepts. We look about us at the biological world in its unbelievable complexity and demand to summarize it in a few words. Living forms, we say, are a result of the interplay of two forces, conservation and mutation. It isn't true—we know it isn't true—but it is difficult to resist vices and we adopt this point of view temporarily to discard it later.

What we mean by conservation is merely the astonishing regularity with which organisms and cells regularly produce upon multiplication similar organisms and cells. On the mechanistic level, conservation is provided for by mitosis; on the molecular level, it is provided for by gene—that is, DNA—replication. Let us discuss gene replication.

REPLICATION

We know that the genetic material of a given cell must be able to replicate itself exactly. The double-stranded nature of DNA provides an attractive hypothesis for the mechanism by which genetic material doubles. Assume that, at replication, the two strands of a molecule of DNA separate, and each single strand maintains its structural integrity. Base pairing between adenine and thymine and between guanine and cytosine then implies that each single strand can force the synthesis of an image of itself in the complementary strand. Take a molecule of DNA and label the strands I and II. Normally, I and II form a tight couple. At replication, I and II separate. Strand I directs the synthesis of a new II; similarly II directs the formation of another I. The result is two molecules of DNA that are identical with the first. Such a method of replication would permit preservation of identity from generation to generation, for a strand once formed would remain intact and would dictate the structure of its newly formed complementary strand. In addition, if the base sequence were to serve as the hereditary code, we have a way to preserve and transmit the code.

The question now is, what evidence is there that replication in the cell involves a process of strand separation and directed synthesis of complementary strands, with each original strand remaining intact? A number of exciting experiments have recently been carried out to examine these points. A series of studies by Matthew Meselson and Frank Stahl will be described at length to illustrate one kind of experimental attack now being made.

If you look at a table of atomic weights, you will see that nitrogen has an atomic weight close to 14. Most nitrogen atoms do have this weight, but there is a stable (nonradioactive) isotope of nitrogen with an additional neutron in its nucleus and an atomic weight of 15. This is heavy nitrogen, N^{15}. Bacterial cells will grow in a medium containing either normal nitrate, $N^{14}O_3^-$, or heavy nitrate, $N^{15}O_3^-$. If the nitrate ion is the only source of nitrogen for the bacterium, the nitrogen atom from NO_3^- will go into all the molecules that have nitrogen as part of their structure—in particular, the bases of DNA. DNA, extracted from bacteria grown in $N^{15}O_3^-$, will have a higher density, weight per unit volume, than DNA from bacteria grown in $N^{14}O_3^-$. This difference in density makes it possible to separate "heavy" DNA from ordinary DNA in the centrifuge. Under standard conditions, movement of a molecule in a centrifugal field (sedimentation) depends on its density. Heavy DNA, with nearly all of its nitrogen as N^{15}, will have a higher sedimentation rate and move a greater distance before coming to rest (equilibrium sedimentation $= S_e$) than will ordinary DNA. A DNA molecule with half N^{15} and half N^{14} will have an S_e just in between the others.

Meselson and his associates grew cells of the bacterium *Escherichia coli* in $N^{15}O_3^-$ for many generations until virtually all of the DNA contained N^{15} (DNA^{15}). The cells were harvested, put into a medium with *only* $N^{14}O_3^-$ and grown for just that length of time required for each cell to divide once. The total number of cells had now doubled, the total amount of DNA had doubled, and each molecule of DNA had doubled. The total amount of DNA^{15} must equal the total amount of normal DNA. A sample of the cells was then taken, the DNA extracted, and S_e measured. It fell just between that of DNA^{15} and normal DNA.

Let us consider the predictions of our original hypothesis on the mechanisms of DNA replication, and the predictions derived from a contradictory hypothesis. The original hypothesis was: If we start with a cell having DNA^{15}, both strands are heavy and we have strands I^{15} and II^{15}. Let this DNA^{15} replicate in $N^{14}O_3{}^-$ medium. Each strand remains intact. The complementary strand synthesized by I^{15} is then II^{14}, and that of II^{15} is I^{14}. The cell divides. One cell gets $I^{15}II^{14}$ DNA, the other $I^{14}II^{15}$ DNA. The DNA molecules are identical; each has half of its nitrogen as N^{14}, the other half as N^{15}, and the S_e of this "hybrid" DNA should be between DNA^{15} and normal DNA.

A contradictory hypothesis is: DNA strands do not remain intact. They separate, break down, and the new strands are formed at random from the N^{15} bases contributed by the original DNA^{15} and the new N^{14} bases made by the cells from the $N^{14}O_3{}^-$ in the medium. True, strand I makes a new II and strand II a new I, but now the N^{15} in each DNA molecule is distributed equally between the strands, and we wind up with two DNA molecules, $I^{\frac{1}{2}N15, \frac{1}{2}N14} II^{\frac{1}{2}N15, \frac{1}{2}N14}$. Remember that before, *one* strand in the DNA molecule had all the N^{15}, as in $I^{15}II^{14}$. The two types of DNA predicted by the two hypotheses cannot be distinguished by S_e measurements.

Note carefully that while the predicted *distribution* of N^{15} is sharply different, both types of molecules have the same number of N^{15} atoms, and it is the *number,* not the distribution, that determines S_e. However, let the bacterial cells go through just one more division in $N^{14}O_3{}^-$. Now the predictions lead to divergent consequences that are experimentally measurable. According to the second hypothesis, again the DNA breaks down. Since each strand has but one-half of its original N^{15}, when it forms a new strand, each new strand will have one-quarter of the original N^{15}. More important, *every* strand will have about the same amount of N^{15}.

Consider the consequences of the original hypothesis. After one division, $I^{15}II^{14}$ and $II^{15}I^{14}$ were formed. $I^{15}II^{14}$, in $N^{14}O_3{}^-$, will now separate and form $I^{15}II^{14}$ and $II^{14}I^{14}$! Instead of all the strands being alike, and having one-quarter of the original N^{15}, we have two *different* species of DNA, one with all N^{14}, normal DNA, the other species still with one-half its original N^{15}. If they exist, they can be separated. The experimental result was that two species of DNA did exist, not one. Moreover, the "heavy" species contained one-half the original N^{15} content, not one-quarter.

This is a beautiful experiment, brilliantly conceived and executed. If you have mastered the concepts, you have a taste of the kind of work going on today in molecular biology. Of course, the experiment does not provide a final answer. It is consistent with, and offers support for, our original hypothesis. It disproves the second hypothesis, but there are other possibilities. Why not think one up, and see if you can plan an experiment to test it? (Suppose DNA is not two-stranded, suppose it is four-stranded?)

There is evidence, then, which points to a semiconservative method of DNA replication *in vivo,* in the living cell; that is, a strand once formed has a touch of immortality.

Given the structural model of coerced replication of DNA *in vivo,* let us now inquire into the biochemical mechanism of DNA synthesis. In an imaginative series of studies, Arthur Kornberg and his collaborators have shown that DNA can be synthesized *in vitro,* i.e., in the test tube. Synthesis requires the enzyme DNA polymerase, the four deoxynucleotides in the form of their triphosphates (deoxyadenylic, deoxyguanylic, deoxycytidylic, and deoxythymidylic acids), and a small amount of DNA itself. This absolute requirement for DNA as a primer for synthesis is entirely consistent with our model of replication!

There is an additional feature of interest in the required presence of DNA as a primer in the synthesis. From extensive physico-chemical studies, it has been learned that when double-stranded DNA is heated to a critical temperature, the two chains separate. The extent of separation, whether complete or partial, depends on experimental conditions, but what is important is that the separated state can be maintained by rapid cooling. Heat-separated DNA (generally called "single-stranded" DNA) is found to be a far better primer than double-stranded DNA. The latter will serve as primer, but this requires the presence of a second enzyme to induce breaks and, thereby, single-stranded ends. The newly-synthesized DNA is of course identical to the primer DNA.

Is it possible, using this system, to synthesize biologically active DNA? Rose Lipman and Waclaw Szybalski, using Kornberg's enzyme, have obtained preliminary evidence that it can be done. They took DNA of a strain of *Bacillus subtilis* that was resistant to the antibiotic streptomycin, made the DNA single-stranded, and from it succeeded in synthesizing more transforming activity in the tube.

We can ask another question. Does a DNA molecule separate into two separate strands and then duplicate, or does it start at one end, and simultaneously unravel and duplicate? The Australian investigator Cairns has obtained some intriguing photographs of the process of replication through the combined use of electron microscopy and autoradiography. (Autoradiography involves getting radioactive atoms into the right structure, then letting these atoms take their own picture on an appropriately sensitized film.) This combination of techniques used by Cairns permits us to take a look at DNA molecules. In Fig. 5-1, you can see his photograph of the

DNA of *E. coli*. It is a picture of the chromosome during replication. This and other photographs suggest that replication does not occur by separation of strands, rather it starts at a point, and the two strands separate in a local region, and are then duplicated. This picture is of interest, as we know from other types of experiments that chromosome replication is a process which starts at a fixed point and proceeds unidirectionally, and in fact, relatively slowly. At the moment we are concerned only with the fact that the strands do not have to separate completely.

Fig. 5-1. Autoradiograph of E. coli B3 following incorporation of tritiated thymidine for a period of one hour (two generations). The arrow shows the point of replication. Exposure time was 61 days. The scale shows 100 microns. (Courtesy Dr. John Cairns.)

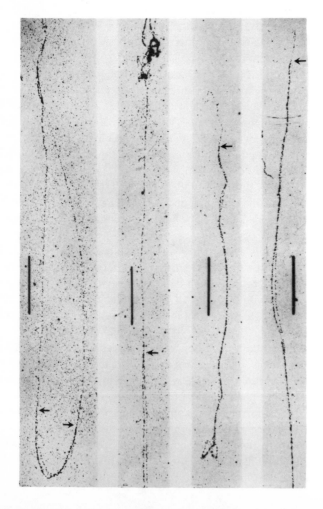

The experimental information discussed above has all come from work with microorganisms. Observations on higher organisms, although fewer and more difficult to obtain experimentally, offer no contradictions. The enzyme described by Kornberg, DNA polymerase, has been found in many cells of higher organisms and is located right in the nucleus where it belongs. As another example, consider the following experiments carried out on chromosomal replication in plants. The investigator, J. Herbert Taylor, used deoxythymidine (one of the four DNA bases) in which some of the hydrogen atoms were replaced by tritium, a radioactive isotope of hydrogen. If root tips of the bean *Vicia faba* are grown in the presence of radioactive thymidine, the thymidine is incorporated into DNA as deoxythymidylic acid and is seen to be distributed throughout all the chromosomes. If the chromosomes are now permitted to divide once in the *absence* of a radioactive label, we now see one strand which is labeled with radioactivity and one unlabeled strand. The level of organization at which these observations are made is at the chromosomal, not the molecular, level. But the observations again point to the conclusion that there is a basic "doubleness" to the replicating genetic structure, be it a visible chromosome as we find in *Vicia faba* or the DNA molecules we find in bacteria.

All the contemporary experimental evidence points to the conclusion, therefore, that genetic material is a double-stranded structure and that this structure replicates in a semiconservative way, i.e., by separation of the two strands, with each single strand maintaining physical integrity during replication.

While studies on the chemical structure, synthesis and mechanism of replication of DNA are going on, other people are worrying about another problem, the problem of the structure of gene information. Genes control the sequence of amino acids in proteins. Genes are DNA. Therefore, DNA controls the sequence of amino acids in proteins. DNA consists of four bases. Proteins consist of 20 amino acids. How can DNA be coded so that its genetic information can be read off as an amino acid sequence? To present this problem clearly, we must first discuss the mechanism of protein biosynthesis.

PROTEIN BIOSYNTHESIS

Like DNA, proteins can also be synthesized in the test tube. Present methods are inefficient as are most miracles, but they tell us a lot. What do we need? Primarily amino acids and RNA; in fact, three different kinds of RNA. The second major polynucleotide of living matter, RNA, exists in all cells. It differs from DNA in its bases, and its pentose (the 5-carbon sugar component), and in its structure which is not uniformly double-stranded. The bases are adenine, guanine, and cytosine as in DNA but with uracil in place of thymine; its pentose is ribose, not deoxyribose.

$$
\begin{array}{c}
\text{OH} \\
|\\
\text{C} \\
N \diagup \quad \diagdown \text{CH} \\
\| \qquad \quad \| \\
\text{C} \diagdown \quad \diagup \text{CH} \\
\text{HO} \qquad \text{N}
\end{array}
\qquad
\begin{array}{c}
\text{OH} \\
|\\
\text{C} \\
N \diagup \quad \diagdown \text{C—CH}_3 \\
\| \qquad \quad \| \\
\text{C} \diagdown \quad \diagup \text{CH} \\
\text{HO} \qquad \text{N}
\end{array}
$$

Uracil   Thymine

$$
\begin{array}{c}
\text{O} \\
\diagup \quad \diagdown \quad \text{CH}_2\text{OH} \\
\text{H} \diagup \qquad \text{C} \diagup \\
\text{C} \qquad \qquad \diagdown \\
\text{HO} \backslash \text{H} \quad \text{H} \diagup \qquad \text{H} \\
\text{C—C} \\
\text{H}^{2'} \ \text{OH}
\end{array}
\qquad
\begin{array}{c}
\text{O} \\
\diagup \quad \diagdown \quad \text{CH}_2\text{OH} \\
\text{H} \diagup \qquad \text{C} \diagup \\
\text{C} \qquad \qquad \diagdown \\
\text{HO} \backslash \text{H} \quad \text{H} \diagup \qquad \text{H} \\
\text{C—C} \\
\text{OH OH}
\end{array}
$$

$2'$-Deoxyribose   Ribose

There are three major types of RNA in the cell, ribosomal RNA, transfer RNA, and messenger RNA. We come to them in turn. Experimental evidence tells us that protein formation occurs mainly in the cytoplasm, in specific particulate elements called microsomes. Microsomes consist predominantly of particles called ribosomes. About one-third of the ribosome is protein and the remainder is large-molecular-weight (about 10^6) RNA called ribosomal RNA. If DNA is in the nucleus, and proteins are made in the cytoplasm, something has to carry the message. This is where another type of RNA comes in, messenger RNA. Lastly, protein synthesis requires amino acids, and a third type of RNA, transfer RNA. Transfer RNA is low-molecular-weight RNA (about 15,000) that combines with amino acids and transports them to the microsome.

Our current picture of how proteins are made is roughly as follows. The gene DNA makes a complementary messenger RNA (m-RNA) which moves to the cytoplasm. m-RNA serves as a tape, on which the various transfer RNA's can attach. Each transfer RNA molecule has a two-fold specificity. It must pick up a specific amino acid, and then lock in at a specific site on the messenger RNA. The ribosomes then attach at one end of the m-RNA tape and read it by moving along. In moving, they knit to form an amino acid sequence, so that with movement, the protein chain grows.

We show in the diagram that a given messenger can have a number of ribosomes (the unit is called a polysome) reading it at one time, and this solves the problem of one messenger serving for the synthesis of many identical protein strands. In fact, Alexander Rich has pictures of this very process. We do not understand how the transfer RNA is fed in, nor what the actual role of the ribosome is. But we do have a sound knowledge of the components required for protein synthesis and a general idea of how they interact. For a more detailed discussion of protein biosynthesis, see the previously mentioned volume by W. D. McElroy in this series.

MESSENGER RNA

In the preceding paragraph, we blandly stated that messenger RNA must exist. If true, this tells us one of the most important facts of genetics, namely that the chemical reaction a gene carries out is the formation of a specific RNA. The evidence is moderately overwhelming. We now know of an enzyme which catalyzes the synthesis of RNA *in vitro,* and which requires DNA as a primer (a DNA-dependent RNA polymerase). If the base composition of this RNA is compared to that of the DNA primer, it is found to be complementary to the DNA. Another clever experimental device serves to emphasize the complementary nature of m-RNA to DNA. If the DNA primer is made single-stranded by heating, then mixed with the newly synthesized RNA, hybrids can form consisting of one strand of RNA and one of DNA, instead of the customary two strands of DNA. These laboratory discoveries are not artifacts. They made it possible to go into the cell to look for RNA complementary to DNA. In cells such as bacteria infected with viruses, rapid protein synthesis is going on, and we find newly synthesized RNA. The newly-made RNA, in contradistinction to the other RNA in the cell, is complementary to the viral DNA by the test of hybrid formation. Thus, the primary reaction carried out by the gene is a reaction involving the formation of a specific RNA. Armed with these additional facts, let us consider the problem of coding.

CODING

The coding problem is easily stated. What we need is a four-letter code (the four nucleotides in DNA) to give a dictionary containing 20 words (the amino acids). Rapid progress has been made in our understanding of the code. But many of our present uncertainties stem from ground rules laid down in our earlier consideration of the problem as one in cryptography, so let's start there.

For ease of discussion let the nucleotides be the letters A, B, C, D. Assume no restrictions on base sequence in DNA. We have, therefore, as many repetitions of each letter as we please, and they can be arranged in and order, for example, ADCCCCBAAADB . . . One letter is obviously not enough to specify one amino acid, for that would take care of only four of the 20 amino acids. A two-letter word is also insufficient. The first letter can be either A, B, C, or D. That gives four possibilities. The second letter of the word can also be either A, B, C, or D. This gives $4 \times 4 = 16$ different words (AA, BC, BD, DB . . .), and again this falls short. We try three-letter words. The first letter can be either A, B, C, or D, as can the second and the third (e.g., ABC, BCA, BDC, CDB, AAA . . .). This gives $4 \times 4 \times 4 = 64$ possible words to specify 20 amino acids, an embarrassment of riches. Too many words means we must have some restrictions on which of the words make sense and which are nonsense, to prevent the amino acids from becoming confused.

First, what about the form of the code? That is, do we want an overlapping code or do we want a nonoverlapping one? For instance, if the word ABC spells amino acid 1, and BCD amino acid 2, at least two arrangements are possible:

A B C D	A B C	B C D
1 2	1	2
Overlapping	Nonoverlapping	

Present data contradicts an overlapping triplet code. If mutation occurs by changing one letter, i.e., mutation of A B C D $\longrightarrow$ A C C D,

$$1 \quad 2 \quad 3 \quad 4$$

one would expect substitution of two amino acids in the protein. This is false as there are many cases of single amino acid substitutions, and no cases of doubles. Another objection is that the use of four letters to form three-letter words make it impossible on an overlapping basis to create a code which does not result in certain amino acids occurring together frequently. Since evidence does not support this, attention was first directed towards nonoverlapping, "comma-less" codes. Formally, such a code must introduce certain restrictions, since overlapping triplets must give "nonsense" information, i.e., a triplet which does not spell an amino acid. Consider the sequence A B C B C D A B C. The four overlapping triplets, BCB, CBC, CDA, and DAB, cannot code, since overlapping is not permitted. This introduces some interesting consequences. As mentioned earlier, four letters, three at a time, give 64 combinations ($4^3 = 64$). Certain combinations are not allowed. AAA is impossible since if we had AAAAAA, the overlapping triplets would code. Thus we know that identical triplets of A, B, C, and D have to be nonsense triplets. This eliminates 4 words and leaves 60 combinations for a comma-less code. The cyclic permutations of any triplet cannot be used; i.e., given a sense triplet ABC, then BCA and CAB must not be able to code. Consider A B C A B C. If ABC is sense, then overlapping makes BCA and CAB

nonsense. Now only one-third of the possible combinations can be sense triplets, the rest must be nonsense. One-third of 60 = 20, the magic number.

This is one of the many dictionaries that can be written:

A	A		A	A		A		A
B				C		B → D → B		
	B		B	B		C		C
				C				D

A typical word is formed by taking any letter on the left, followed by the center letter, then any letter on the right. As the arrows indicate, ADB is a sense word. All words not in the dictionary are nonsense words. For example, ABA and ABB are in; ABC and ABD are out.

The surprising fact emerges that a comma-less code can be constructed, using four bases, three at a time. The four bases of DNA, used as triplets, could be readily arranged to code the amino acid sequence of a protein.

The question is—does such a code get used in real life? Is, for example, the bacterium *E. coli* sufficiently moved to make use of it? The answer is no. A pity, but it is not man's art but nature's that concerns us. Why are we so sure this code is not used? The answer comes from the test tube.

A problem now facing biochemists is how to isolate one message. It is possible to separate messenger RNA from both ribosomal and transfer RNA, but in the messenger fractions how do we isolate just one message? This isn't easy, and in fact it hasn't been done, but one can resort to a bit of chicanery. For instance, we mentioned earlier that TMV consisted of RNA and protein, and that the RNA was infective. On inoculation into tobacco leaves RNA gives rise to more RNA, more protein and finally TMV, i.e., the RNA carries the message about TMV protein. Will TMV-RNA serve as a messenger for an *E. coli* enzyme-forming system? Marshall Nirenberg and J. H. Matthei tried this experiment and it worked. This led them to try a second experiment. If *E. coli* can use TMV-RNA, as a messenger RNA, what about synthesizing a message? It is possible in the laboratory to synthesize RNA of known composition, by using the enzyme RNA-polymerase. Suppose we do so and use this as m-RNA in our protein-synthesizing system. Will it make a protein? Suppose a protein is formed; since protein synthesis is determined by m-RNA which, in turn, is only a reflection of the DNA, knowledge of how m-RNA codes will give us the answer to the DNA code. In 1960, Nirenberg and Matthei did this experiment. With the enzyme RNA-polymerase and the nucleotide uridine triphosphate, they synthesized an RNA polymer with just one base, polyuridine (poly-U). They then extracted from E. coli all the other ingredients required for protein synthesis, added poly-U, and synthesized polyphenylalanine, a protein with one amino acid. This experiment is recorded in history as the U-3 incident. The U-3 experiment gave us our first word: UUU spells phenylalanine. Fantastic.

The U-3 incident also set off a coding war after the realization that with one word known it was now possible to get other words. It also told us, of course, that some of the ground rules laid down earlier would have to change, for as you recall, triplets of our base had been ruled out as giving rise to overlap. This means we must read the code some other way, as the actual code would seem to tolerate overlapping triplets. The scientists Crick and Brenner came through. They suggested overlap is no problem if a device is used that reads three bases at a time (a reading frame) and starts at a precise point of origin. They found the suggestion so provoking that they went out and got some exciting data to support it. But what about nonsense triplets? Do we have to devise a code of only 20 words, or may we use a degenerate code, one in which a given amino acid is coded for by more than one triplet? This is today's problem, and we can summarize where we now stand.

The fact that poly-U will stimulate polyphenylalanine formation suggests that use of synthetic polymers of varying compositions can give us still other words. They have, and by this means a number of dictionaries have been proposed by Nirenberg and, independently, Ochoa. Their work tells us of specific triplets associated with specific amino acids, but their investigations also tell us that the code is degenerate.

There is a second line of inquiry which also gives clues to the actual dictionary. This concerns amino acid replacements found in proteins formed by mutants. As we discussed in the preceding chapter, it can be shown that as a consequence of mutation, glycine in E. coli t'ase is replaced by arginine. We also know the same glycine can be replaced by glutamic acid. The triplet coding glycine must be related to the triplet coding arginine, and to the one coding glutamic acid. We have a large amount of amino acid replacement data obtained from the study of human hemoglobins, from the mutations in the enzyme t'ase, and from mutations of the protein coat of TMV. If these data are pooled with the synthetic polymer data we come up with a code suggested by Thomas H. Jukes (there are others) and given in Table 5-1. We present this simply as an indication of the type of dictionary we are thinking about. But while we now have words, we have still to puzzle over problems of nonsense, degeneracy, redundancy, and reading frames. For the fact is that the genetic code does not conform to the rigorous standards set by orthodox cryptography.

Table 5-1

SUGGESTED ASSIGNMENTS OF TRIPLETS
TO CODING FUNCTIONS OF THE AMINO ACID CODE

Amino Acid	"Modified" Base Pairs[a]	Suggested Functional Triplets		
		* = U	* = A	* = C
Lysine	A*A	AUA	AAA	
Asparagine	*AA,C*A	UAA,CUA	CAA	
Histidine	A*C	AUC		ACC
Glutamine	*AC,GG*	UAC	AAC,GGA	
Glutamic acid	A*G	AUG	AAG	
Tyrosine	A*U	AUU		ACU
Threonine	*CA,*GC	UCA	ACA	CCA,CGC
Proline	C*C	CUC	CAC	CCC
Alanine	C*G	CUG	CAG	CCG
Serine	C*U,*CC,*CG	CUU,UCC	ACG	
Aspartic acid	G*A	GUA		GCA
Methionine	*GA	UGA		
Arginine	G*C,G*A	GUC	GAA	GCC
Glycine	G*G	GUG	GAG	GCG
Tryptophan	*GG	UGG		
Cysteine	G*U	GUU		
Isoleucine	*UA,*AU	UUA	AAU	CAU
Valine	U*G	UUG		
Leucine	U*C,*CU,*AU,*GU	UUC,UAU,UGU		CCU
Phenylalanine	U*U	UUU		UCU

After Thomas H. Jukes, in *Informational Macromolecules*, edited by Henry J. Vogel, Vernon Bryson, and J. Oliver Lampen. New York: Academic Press, 1963.

[a] The doublets in this column are considered inviolate; substitution of the star by U, A, or C provides for degeneracy as seen in columns 3 to 5.

Before considering these problems, let us take a closer look at mutation in the light of what is known about the DNA code.

MUTATION

Gene mutation is any alteration in the nucleotide sequence of a gene, but a mutation may also occur by a change in the geometry of the chromosome. Many types of alteration are possible. Mutation may occur by deletion, i.e., by physical loss of a genetic region. Regard a chromosome:

If region A were lost, any traits carried in that region would be lost. Organisms with deletions are characteristically unable to undergo "reverse mutation"—mutation back to the original phenotype. Mutation by transposition of portions of the chromosome is also known to occur. For instance, if region A were moved to the end of the chromosome or to another chromosome, it could lead to alteration in function. Cases of this type are well known in corn and in fruit flies.

The most commonly occurring mutations are point mutations. We can account for point mutations as changes of nucleotide sequence in the DNA which alters the triplet code and forces consequent changes in the nucleotide sequence of RNA and the amino acid sequences of proteins. Amino acid replacement data obtained by Charles Yanofsky with the A t'ase protein of *E. coli* will serve admirably to illustrate the amino acid changes. Assume, as in the diagram, that the triplet GUG codes glycine, GUC arginine and AUG glutamic acid (see Table 5-1):

$$\text{GGG} = ?$$
$$\updownarrow$$
$$\underset{\substack{\text{Glutamic}\\\text{acid}}}{\text{AUG}} \leftrightarrow \underset{\text{Glycine}}{\text{GUG}} \leftrightarrow \underset{\text{Arginine}}{\text{GUC}}$$

GUG → GUC would change glycine to arginine, GUG → AUG changes glycine to glutamic acid. Both these single amino acid substitutions are, of course, known. Let us consider a change of GUG → GGG. Assume that GGG is a nonsense triplet, a triplet that codes no amino acid. Then mutation GUG → GGG could lead to loss of ability to form enzyme. If the code is completely degenerate, base substitution can have only one effect, namely missense mutations—triplets that code a different amino acid. If the code does have nonsense triplets, then base substitution offers two general consequences, nonsense and missense mutations.

Missense mutations are known but the existence of nonsense mutations is hard to prove. It is difficult to prove something is not made if you have no way of recognizing what should be made. Point mutations are known which result in the loss of a detectable product. Whether these are nonsense or missense is still being examined. In any event it is clear that we would like to have, as mutagens, reagents that cause base substitutions.

Ionizing radiations, such as X-rays, are highly mutagenic. The effects, in part, are chromosomal and are due to deletions, inversions, and translocations, the latter two being rearrangements of chromosomal material. These occur because X-rays can break chromosomes, some pieces can be lost, while others can recombine but in different order. X-rays also probably cause base substitution, but this is a point that is difficult to prove. Ultraviolet light is also highly mutagenic. Like X-rays, ultraviolet radiation can give rise to both chromosomal alterations and point mutations. Again, the molecular basis of the action of ultraviolet light is not clearly known.

In recent years, another class of mutagens has appeared and has been found to be of great interest in connection with base substitution and mutation. This is the class of chemical mutagens, one of which is a common inorganic acid, nitrous acid (HNO_2). HNO_2 is a potent oxidizing agent. It can react with primary amino groups, such as are present in the bases, and replace the amino group with an oxygen atom. The reaction of HNO_2 with cytosine yields uracil. HNO_2 can also react with adenine and convert it to a different base, hypoxanthine. Suppose HNO_2 does react with cytosine in a DNA molecule and converts it to uracil. The strand carrying uracil, upon replication, might now make a mistake in pairing, and, rather than pair with guanine, pair with adenine. In a subsequent replication, adenine, in turn, will pair with thymine. HNO_2 will thereby have effected the transition of a guanine-cytosine pair to an adenine-thymine pair, thus altering the base sequence of the DNA.

Nitrous acid, however, does not offer the specificity that one might like, for it will react with the amino group in three bases. Recently another class of mutagens has appeared—base analogues. A base analogue is a compound with properties that are similar but not identical to those of the natural bases. A good example is 5-bromouracil (BU). When BU is offered to a cell such as *E. coli,* the unsuspecting bacterium reads it is thymine and puts it into its DNA. Nearly 70–80 per cent of the thymine in the *E. coli* DNA can be replaced by BU without impairing its growth, but consider the consequences of division. Thymine normally pairs with adenine but BU can pair with both adenine and guanine. Suppose we start with the original DNA (see Fig. 5-2). On replication the guanine in the DNA is looking for cytosine but it can occasionally be hoodwinked into taking BU. Once BU gets into the complementary strand, on further replication it, in turn, will preferentially pair with adenine. Adenine, at the next replication,

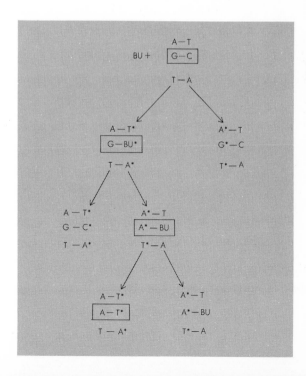

Fig. 5-2. Induction of G/C → A/T transition by 5-bromouracil. For details see text. (* = bases of new strands.)

will pair with thymine. The net effect of this incorporation error, as shown in the chart, is a "transition" of a G/C pair to an A/T pair.

A second base analogue, 6-aminopurine (AP) has a related effect. AP is incorporated in place of adenine and, as a consequence of the incorporation error, appears to cause transitions of A/T to G/C. The combined use of these base analogues in the hands of Benzer has already begun to give base sequences of genes in terms of A/T and G/C pairs. In time we may have a specific reagent for each base which may then permit the establishment of the base sequence of a given gene. Certainly this is a promising approach.

To sum up, we have discussed some aspects of the chemistry of the genetic material, DNA. We have gone over some of the current thoughts on the nature of the code which directs the formation of a specific protein. And we have explored one suggestion on the way genetic information is communicated to the appropriate centers of synthesis. From these considerations, we go to the problem of the molecular basis of recombination.

The Molecular
Structure
of a Gene

CHAPTER SIX

In an earlier chapter, we introduced briefly the phenomenon of genetic recombination. In terms of inheritance, recombination refers to the appearance of new combinations of traits in progeny cells or organisms. These new combinations are formed either as a consequence of random assortment of chromosomes during meiosis or of genetic exchange produced by crossovers. Random assortment is straightforward, but how can crossing-over be explained? In the preceding chapter, we discussed how knowledge of the chemistry and structure of DNA suggested many of our ideas and theories of replication and coding. What insight does this knowledge give us into crossing-over? The problem is of interest since if we are to map the fine structure of our gene, it wouldn't be a bad idea to understand what it is we are mapping.

BREAKAGE-FUSION

In meiosis, prior to the first division, intimate pairing of the chromosome strands (chromatids) takes place. Breaks in the strands can then occur, and by reciprocal exchange chromatid segments can be recombined. This, of course, permits the recombination of linked genes. These events have been seen microscopically, and are entirely consistent with the results of the classical inheritance test.

Although we know that breakage-fusion recombination occurs and although we can see it, the mechanism is unknown. This reflects our ignorance about chromosomal organization in general, and specifically our ignorance of how the DNA and protein components are arranged. However, while the problem of chromosomal structure is largely obscure, we are acquiring a picture of a gene, a picture built largely from *intragenic* recombination data.

It was long thought that breakage-fusion recombination would occur only between genetic regions controlling different traits. In large part, this conclusion still holds. It is probable that breakage-fusion recombination does occur primarily (or most frequently) between genes, not within a gene. However, recombination within a gene does occur. What, then, is the nature of this intragenic recombination? Is it breakage-fusion recombination? If it is, how do the deoxyribosephosphate backbones of the DNA chains ever get close enough to enable switching of exact and identical nucleotide sequences of both strands? Phosphate groups are all negatively charged under the normal physiological conditions of the cell. If DNA strands did pair closely enough to exchange bonds, strong forces of repulsion would have to be overcome. Additionally, the phosphate ester bond is a relatively stable bond, and considerable energy would have to be mobilized to break and reform these. The benefits of sex, i.e., recombination, are certainly sufficient for the cell to call on all of its resources to do this. However, these difficulties have led to consideration of an alternative mechanism, copy-choice.

Copy-choice is assumed to work in the following manner. DNA replicates as single-stranded DNA, with each strand prescribing its complementary strand. An error in replication could occur as is shown in Fig. 6-1. Suppose a strand complementary to strand A were in the process of replication. Assume that another strand, identical to A except for an alternative allele (strand B) were also in the cell and close to the newly forming complementary strand. The replicating strand on rare occasions might

Fig. 6-1. Copy-choice recombination. The two parental strands are homologous but not identical.

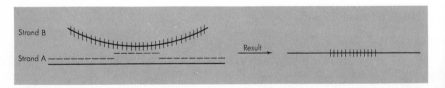

switch and use strand B as its model for a small region. The replicate, then, would not mirror the parental strand, but would effectively recombine the genetic codes of the two strands. With this model, there is no need for breakage-fusion or the formation of reciprocal classes.

Recently, however, experimental evidence indicates that transforming DNA does not offer the replicating mechanism a choice, but is physically incorporated into the replicating DNA strand of the recipient parent cell. In addition, many of us, in reviewing the anomalies of intragenic recombination, feel that at present there is no rational answer. Copy-choice will probably soon fade from consideration, for there is increasing evidence of DNA interaction and physical incorporation. Intragenic recombination is still a fascinating puzzle, and its clarification at the molecular level will be rewarding. You may guess that there are still a few things for you to do.

FINE STRUCTURE

What we have just said is that alleles within a gene can be mapped, even if we don't understand the underlying mechanism. Take our favorite example, the allelic series of tryptophan synthetase mutants discussed in Chapter Four. These are mutations of a single gene on chromosome-2 of neurospora. These allelic strains, if crossed with one another, do yield infrequent tryptophan-independent progeny, and we can now map the genetic region by determining the frequency with which different allelic crosses give rise to tryptophan-independent progeny. The map can be prepared using this one criterion, and the map is consistent as map distances are additive. Similar maps of genes controlling other activities have been prepared for a number of genes in fungi, bacteria, and bacterial viruses, such as the r-2 gene of the T4 bacteriophage owned by Seymour Benzer, and the t'ase genes of *E. coli* drawn with a fine genetic hand by Charles Yanofsky and his colleagues. These maps are of interest as they give insight into base sequence and illuminate 3-dimensional problems of gene products.

All maps are similar in showing a large multiplicity of mutational sites. Such mutational subdivisions of a functional area are called mutons. A muton, according to the classical definition of Benzer, is the smallest element that when altered can give rise to a mutant form. A single functional area, a gene, consists of many mutons. Since the gene is DNA, it is reasonable to inquire what the muton represents in terms of nucleotides. A muton should represent the minimum number of nucleotides that must be altered to permit amino acid alteration. Our best estimate comes from Yanofsky. He has observed a mutation in the A t'ase gene which results in the replacement of glycine by arginine (arg+). A second mutation results in replacement of the same glycine by glutamic acid (glut+). A cross of these two mutants (arg+ × glut+) does give recombinants, although at very low frequency; thus these mutants must involve different nucleotides but within the same coding triplet. The recombination frequency is that predicted for recombination between two adjacent nucleotides. All of our present estimates suggest that a muton is a single nucleotide. This we believe must be listed as fact—that alteration of one nucleotide pair will produce mutation.

What about the structure of a gene in terms of recombination? From the experimental evidence just discussed, showing that recombination occurs within a coding triplet, it is difficult to escape the conclusion that the unit of recombination is also a single nucleotide pair.

We now turn to the structure of a gene in terms of function. Here business picks up, for this area is not clearly understood. We must solve two different, though related problems. First, what does a gene look like if one prepares a "saturation" map by mutation, i.e., locates every nucleotide in the gene by mutation? Is the base sequence of a gene uniformly susceptible to mutation and does mutation anywhere along the line result in similar functional changes? Second, must all of the DNA sequence specifying a given enzyme be in a single DNA molecule, or can it be distributed between two molecules? The answer to the first will help shape the answer to the second.

In terms of a saturation map, the r-2 locus which determines plaque morphology in bacteriophage has enjoyed the best success, but this locus unfortunately has no detectable end product to play with and so gives limited answers. Benzer finds great unevenness in this locus, i.e., hot spots and cold spots. By this we mean certain nucleotides repeatedly mutate, others rarely do, and some apparently never. This tells us that for some unexplained reason, nucleotide pairs have unequal stability to a given mutagen. This may not be too hard to live with since DNA has four nucleotides and long runs of one of them, say adenine, would give different local properties than long runs of cytosine. Sequence may have an important bearing on susceptibility of a region to a given mutagenic agent. The problem, however, is a bit more complicated. Let's look at t'ase, for this system answers still other questions.

Certain regions of the t'ase gene appear susceptible to mutation ("noisy" regions), other regions are impervious ("silent" regions). It turns out that this classification does not really reflect absolute differences in susceptibility to mutation. Rather, in order for us to see a mutation, the amino acid substitution has to result in an alteration of function. It sounds like obfuscation, but read on. You will understand.

As we discussed in Chapter Four, we know that many functional differences can be observed between members of the neurospora t'ase allelic series. We can set up a broad classification of these mutants as in Fig. 6-2. Mutation may give rise to either a CRM$^+$ or CRM$^-$ strain. The CRM$^+$'s, as indicated in the chart, can be subdivided into three classes, based on residual enzyme activity: (I) strains which have lost catalytic activity for all three reactions catalyzed by the parental strain; (II) strains without catalytic activity for reaction 1 and 2 but with activity for reaction 3; (III) strains without activity for reactions 1 and 3, but with activity for reaction 2. Class II can be subdivided further in that some of these mutants have no cofactor requirement for reaction 3, others require, in contrast to the parental strain, pyridoxal phosphate, and still others require both B_6 and the amino acid serine.

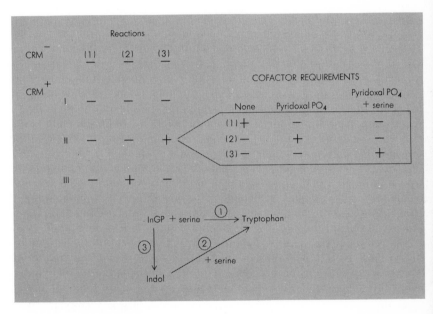

Fig. 6-2. Classes of t'ase mutants.

The enzymatic activity of Class II mutants means that they can still combine with indole and serine but not with InGP, while Class III mutants can still combine with InGP but not with serine.

On mapping these CRM+ strains, we find that alleles that are functionally similar are not scattered throughout the genetic area but are clustered. Mutations of Class I appear to be clustered in a small region near the middle. Mutations of Class II are clustered on the right side, and Class III on the left (see Fig. 6-3).

What does this clustering mean? Remember we are mapping CRM+ strains, i.e., strains which still form a protein much like the parental protein. The amino acid substitution apparently has not caused a major structural

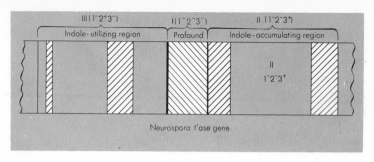

Fig. 6-3. Schematic representation of the genetic fine structure of the t'ase gene. For class type see Fig. 6-2. Hatched areas represent regions in which CRM+ mutations have been detected.

change. If the mutant product protein acts as an enzyme, it must still be capable of interacting with a substrate. Class III mutants cannot react with InGP but this loss has not interfered with their ability to combine with indole and serine. Class II mutants can't react with B_6 and serine, but they can still combine with InGP and convert it to indole. Class I we shall explain shortly.

What we want to emphasize is that what appears "silent" and "noisy" need not reflect the *occurrence* of a mutation. Rather, amino acid substitution in certain areas can alter catalytic activity of the enzyme while in other areas it does not. Remember that for a mutation to be detected, it must have a deleterious effect on the ability of the enzyme to combine with one of its substrates. All this is consistent with what we know about enzymes. Some of the amino acids are directly involved in the catalytic sites, others are not. It is to be expected, therefore, that mutations leading to amino acid substitution will differ sharply in effect depending on the nature of the substitution and where it takes place.

Some interesting support for this concept comes from the superb structural analysis of the protein myoglobin carried out by Kendrew and Perutz. By X-ray crystallographic diffraction and brilliant imagination, they determined the entire 3-dimensional structure of myoglobin. One fact emerges clearly. The protein molecule is folded in such a manner as to bring into close spatial relationship amino acid groups relatively distant from one another in the chain. Thus, we find areas in which folds occur as well as areas of straight protein chain. The importance of these structural considerations becomes immediately apparent, since there is a great deal of evidence available that permits the extension of the picture of myoglobin to other proteins. If portions of neighboring chains of amino acids brought together by folding constitute sites of enzyme activity, and other portions of the molecule represent regions of structural support, then a reasonable correspondence can be set up on the one hand between the noisy regions of the gene and the sites of enzymatic activity, and on the other hand, between the quiet regions and the areas of structural support. This idea has found support from recent data obtained with *E. coli* t'ase in which it has been shown that some amino acid substitutions yield no detectable enzymatic differences. All this suggests that a gene has a built-in safety factor in terms of mutation, for approximately 75 per cent of the neurospora t'ase gene is "silent."

We come now to our second problem. Must all the nucleotides of a single genetic area be bound together in a prescribed order for a specific enzyme to be formed? This has been studied by making use of the cis/trans test, a test which gave rise to the unit of function, the *cistron*. The principle of this test is illustrated in Fig. 6-4.

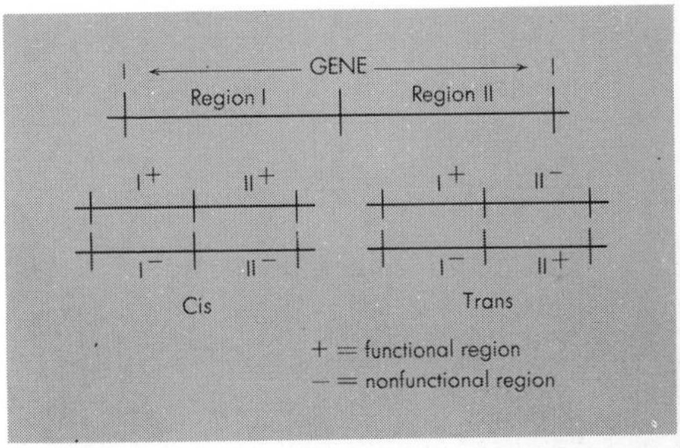

Fig. 6-4. The gene is arbitrarily divided into 2 regions, I and II. The same genetic elements are present in both cases, but are differently distributed.

The problem is, does enzyme formation require the cis configuration shown in the diagram or will the trans configuration work? Surprisingly, the trans position does work, but not for all subdivisions of the gene. The subdivisions that do work in this test, the "functional" subdivisions, are called cistrons. The cistron, like the muton, will disappear with increased understanding of protein structure and synthesis. At present it turns out that a unit of function is exceedingly difficult to define rigorously due to problems of protein-protein interaction.

Protein-protein interactions have been known for many years, but their importance to biology is only now being investigated. We now realize that most enzymes consist of two or three unlike protein chains. They may be dissociable chains, or they may be covalently bound. Additionally, they may also exist as polymers, and the activity of a monomer may not be identical with that of a dimer. To illustrate the main problem, consider the t'ase of neurospora and *E. coli*. In neurospora, this enzyme appears to be a single protein molecule of molecular weight 105,000. We find a single genetic area in which mutations can occur and affect this enzyme. In *E. coli*, however, t'ase is formed of two dissociable protein chains, an A protein and B protein, which together have a molecular weight of circa 105,000. Maximal catalytic efficiency requires the complex. The formation of each protein is controlled by a specific gene, the A protein by the A gene, and B by a B gene. The two genes are closely linked.

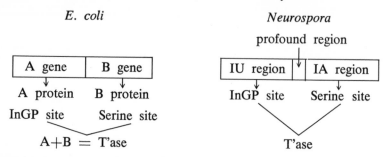

Detailed study of the neurospora and *E. coli* enzymes tells us that they are closely related, the neurospora enzyme probably having evolved from the *E. coli* enzyme. *E. coli* has an A gene and A protein, and this protein has the InGP-combining site. The neurospora gene has the equivalent of the A region, the indole-utilizing region which carries the InGP site. The same relationship is found in terms of a B protein and B region. The neurospora gene, however, has a profound region (Class I mutants) lacking in *E. coli*, and this region is a nucleotide sequence which holds the A and B regions together. This is not surprising, as it is perhaps more efficient to make one protein rather than to have two proteins which are required to search each other out.

A cis/trans test in *E. coli* suggests each gene of t'ase is a functional unit. Mutant A plus normal B gives reaction 2. Normal A plus mutant B gives reaction 3. Mutant A plus mutant B gives no activity.

Now we try a cis/trans test in neurospora. This is done by heterokaryon formation. If two strains of neurospora of similar mating type are grown together, fusion of hyphae between the strains can occur and new hyphae will be formed that carry the two parental nuclei in a common cytoplasm. The nuclei do not fuse to form the heterozygote, but the presence of the two nuclear types in a common cytoplasm does form an off-beat heterozygote. A cis/trans test can then be carried out in that we can determine whether a heterokaryon formed between two allelic t'ase-less mutants is capable of growing on a medium lacking tryptophan.

Certain combinations of alleles do exactly this. The hetrokaryon can grow in the absence of tryptophan and can form a t'ase. However, we find that only alleles of different function show this "complementation," i.e., indole utilizers will complement with indole accumulators, but not with other indole utilizers. The neurospora gene, then, consists of two cistrons. The *E. coli* system, by similar definition, consists of two genes. The terms gene and cistron are obviously closely related and intimately tied to the nature of the functional unit of a protein. Proteins consist of separate functional areas that may be formed as several dissociable chains, or as several chains held together, or as a single chain. Functional proteins may also be formed of combinations of identical subunits and the activity of a dimer may not be identical to the monomer. Thus tests of function are complicated. A cistron may well reflect a functional genetic area corresponding to a separate polypeptide chain. Whether a system is composed of 2 genes, or 1 gene and 2 cistrons, would depend on how the polypeptide units in the enzyme are linked. It is clear, however, that for maximal efficiency, the formation of single protein chains require single messenger RNA molecules, and these in turn are formed from single DNA molecules.

Our discussion so far has dealt with CRM+ mutations, but as you may recall, we also find about an equal number of CRM− in neurospora. Are these different? While some CRM− mutants are deletions most of them are revertible point mutants. Unlike CRM+, CRM− map apparently as point mutations at random throughout the genes. They must, therefore, arise as a consequence of single base changes. These CRM− mutants bring us back to our old friend, the code.

A ready explanation for CRM− would be nonsense triplets. That is, they would represent a base substitution that yields a triplet which does not code for an amino acid. The process of enzyme formation should then stop with the formation of a fragment. The reversion characteristics of CRM− fit this explanation. However, while nonsense mutations are consistent with reversion, the protein products of CRM− do not look like fragments, and a nonsense triplet doesn't account for certain observed curious recombination characteristics.

CRM− mutations can, however, be accounted for in another way. Crick and Brenner have suggested that in reading the code, we could use a reading frame, i.e., the m-RNA has a code that spells "start," and the ribosomes will attach there, and automatically read three bases at a time. Consider the sequence

$$O \quad \overline{ABCABCABCABCABCABCABC} - - - - -$$
$$\underbrace{}_{aa_1} \underbrace{}_{aa_1} \underbrace{}_{aa_1} \underbrace{}_{aa_1} \underbrace{}_{aa_1} \underbrace{}_{aa_1} \underbrace{}_{aa_1}$$

The polypeptide normally coded by this gene would have seven aa_1 at the beginning of the chain. Now what would happen if one base were lost or inserted, e.g., we *lose* the third C:

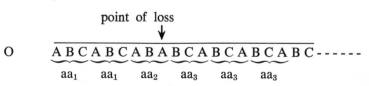

$$\text{point of loss}$$
$$\downarrow$$
$$O \quad \overline{ABCABCABABCABCABCABC} - - - - - -$$
$$\underbrace{}_{aa_1} \underbrace{}_{aa_1} \underbrace{}_{aa_2} \underbrace{}_{aa_3} \underbrace{}_{aa_3} \underbrace{}_{aa_3}$$

We now read correctly to the point of loss, but then start misreading. If the code is degenerate we would get the unrelated product shown above. How could we correct it? The answer—add a base.

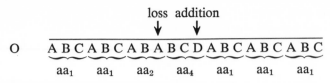

Here we see it would start out correctly, misread a bit, and then start in correctly again. If the code is degenerate, either base addition or deletion alone could result in the formation of products of entirely wrong sequence. A combination of the two would result in a product closely akin to the original. Mutations of this type appear to be produced in bacteriophage by the dye acridine and can be spotted by the ingenious combined use of loss and gain mutations. Intragenic analysis confirms the restoration of a mutation to wild type by the combination of a loss and gain mutation, both of which map at different sites. CRM⁻ mutations could be of this sort, and of interest is that base addition or deletion would account for their recombination characteristics.

CRM⁻ mutants therefore remain alluring, but their explanation must await the next edition. They highlight problems that now confront us. Is the code degenerate? Are there nonsense triplets? What is the reading frame? And how do we code for start and stop? These are the problems we drift on to, as the problems we were concerned about earlier are either resolved, or are found to be irrelevant. As we said earlier, the answer to one question shapes another.

SUMMARY

We can now arrive at the following conclusion concerning a gene. A gene is the unit of DNA in the chromosome in which mutation can occur and affect the ability of the organism to form a specific product. The basic unit of the gene is a nucleotide pair, as this is the basic unit of mutation and recombination. The gene may consist of more than one area of functional activity. In the years ahead, it is expected that the definition of a gene will continue to change, for the gene is an operational unit. In the hands of earlier workers, it was a unit of segregation and, therefore, a unit of recombination. At present, it is a unit of function of DNA, and in the future it may well become a broader unit of function, a unit of development.

Although DNA is the substance that transmits hereditary information, we have skirted the problem of its organization in chromosomes. The visible chromosome of higher forms probably contains the bulk of the DNA of the cell. The inheritance test clearly indicates that genes are arranged in a linear array in the chromosomes of all organisms from bacteria to man. However, the linear array can obviously take many different forms. We know that the DNA, in some way, must be condensed in chromosomes, since the total length of DNA calculated from the Watson-Crick model far exceeds the actual chromosome length. It has been suggested that the DNA is arranged in Christmas-tree style with the DNA as limbs along a protein trunk. Reciprocal recombination then could occur along the trunk, and nonreciprocal recombination along the limbs. This, however, is just speculation. The hard facts are still to be discovered.

Genes
and Development

CHAPTER SEVEN

Genetic researchers during the past ten to fifteen years have been engrossed in the identification of the material basis of the gene, DNA, and of the mechanism whereby the gene functions in the living organism. In our description of some of the current research techniques, of the success that has been achieved, and the questions that arise with each new bit of information, a certain amount of bias inevitably has been introduced. We have implied, at least by emphasis, that nuclear inheritance is the sole mechanism operative in the transmission of heritable traits and that the gene functions by the determination of enzyme structure. If this be gospel, then it must be reconciled with all the facts of biology. Reconciliation is not easy.

Consider but one example. Man starts from a fertilized egg cell, one cell which divides mitotically to produce two cells; these divide mitotically to form four cells—and so on up to a few billion cells, all generated by mitosis, all with a gene content presumably identical with that of the first cell, the zygote. But we know that these billions of cells are not identical in form, function, and biochemistry; they are liver cells, nerve cells, blood cells, eye cells, etc. Differentiation has occurred. We wonder how these differences are generated, how the cells take up their roles in such a precise and orderly fashion. We wonder, in short, what it is that regulates and how regulation is carried out. Is this an additional and exclusive property of the nucleus, or is it an interaction between nucleus and cytoplasm, or cytoplasm and environment? These are problems basic to an understanding of development.

ACQUIRED TRAITS

Heredity has long been thought to be uniquely associated with the nucleus. Mendel's laws, in effect, deny the inheritance of acquired characteristics except insofar as the genes themselves are directly affected. The role of environment in heredity has been a subject of speculation for many years. At the turn of the century, a great deal of discussion took place about the relative roles of environment and cellular factors in heredity. People were aware of the subtlety with which living forms adapted to their environment: the exquisite camouflage of certain insects, birds, and arctic animals, the capacity of deep-sea animals to withstand crushing pressures, the enormous variety of habitats in which bacteria were found, in sulfur wells, in ocean bottoms, in the roots of certain plants, in hot springs. Biological history, in general, indicated that better-adapted species evolved for a given environment. Did the environment play a direct role in this process by directing a specific change in the hereditary apparatus of the organism, or was this caused by random mutation followed by environmental selection? The answer has come from definitive experiments with bacteria and antibiotics.

If you grow a large population of *E. coli* in the laboratory, you will generally find that the population can be destroyed by streptomycin. By growing very large populations in the presence of streptomycin, you can obtain a population resistant to streptomycin. The progeny of these resistant bacteria are found to be resistant to streptomycin even when the antibiotic is removed from the growth medium. The question is, did streptomycin cause the resistance, or did it kill the sensitive bacteria and permit the growth of pre-existing resistant bacteria which arose by mutation independent of the presence of the antibiotic? An ingenious experiment by Joshua Lederberg provided a solution. He devised a method to show that in a large population of *sensitive* bacteria which had never been exposed to streptomycin, about one in ten million bacteria was resistant to streptomycin. He could pick out this bacterium and let it divide and form a very large population, and the entire population was *resistant* to streptomycin—all without the bacteria ever being exposed to the antibiotic.

This experiment, and many others, implies that living things, through gene mutation, are endowed with a certain potential for adapting to a new environment. But this potential is not infinite. Dinosaurs were displaced by other living forms, presumably because they could not meet the challenge of changing environmental conditions. The evidence today, then, suggests that with random mutation, plus the endless reshufflings brought about by gene recombination, living forms can offer a wide variety of response to the environment. The concept that the environment can induce a specific heritable change is not supported by experimental data.

Russian geneticists, however, raised the problem anew. During the Stalinist era, political philosophy in Russia demanded that the environment play a hereditary role, for Marxist dogma was interpreted as decreeing that social forces are capable of forging a fitter race. This doctrine was purported to have received experimental support from the work of Lysenko and his school, who attempted to prove that permanent hereditary changes could be induced in plants by grafting, or in animals and birds by means of blood transfusion. Numerous attempts have been made in other countries to duplicate at least the major experiments on which their conclusions rest, but without success. At the present time, therefore, we can only state that Lysenkoism has proven largely nonproductive in terms of genetic theory; it was essentially a revival of an outmoded genetic theory reapplied for political rather than scientific purposes.

Although experimental data do not support the theory that acquired traits are hereditarily transmitted, how do we classify such phenomena as bacterial transformation and transduction, for here we know that new genetic traits can be acquired by a microorganism from its surrounding medium? In these instances, however, the traits have been acquired by the physical incorporation of genetic material, DNA, and thus the mechanism differs from the older concepts of how acquired traits are transmitted.

Although we may conclude that traits acquired from the environment may seldom become permanently transmissible traits, this does not answer the question about the uniqueness of the nucleus in the transmission of hereditary information. Are there elements in the cell for whose transmission the nucleus is insufficient?

CYTOPLASMIC INHERITANCE

There is a rather easy test for nonnuclear inheritance. In diploid genetics, we know that the egg contributes the bulk of the cytoplasm to the cell formed by gametic fusion. The sperm contributes relatively little. If there are elements in the cytoplasm that require the transmission of a cytoplasmic element rather than the nucleus, we would guess that such traits would be maternally transmitted. The progeny should then show only the maternal trait. Many instances of this type of inheritance are known. A good example can be readily observed in plant variegation.

Variegated plants are those that have areas of pale green, white, or other colors in otherwise normally green leaves. The variegation can be symmetrical or it may be irregular. In Mirabilis, the plant commonly called the four-o'clock, the progeny of a cross between a pale and green plant is pale *or* green depending on the color of the maternal line; i.e., if the maternal line is pale, the progeny are pale; if the maternal line is green, the progeny are green. This is in contradiction to Mendelian genetics. Straightforward nuclear inheritance requires that the progeny of a cross between true-breeding pale and green plants be the same regardless of the maternal trait. The simplest way to account for this type of inheritance is to postulate that a nonnuclear element derived from the maternal plant determines this trait, and we find this to be correct.

The cytoplasmic elements concerned in variegation are the chloroplasts, the bodies carrying the photosynthetic green pigment, chlorophyll. The inheritance pattern of variegation suggests that the inheritance of a chloroplast depends on the transmission of a chloroplast itself. The data, in fact, indicate that if a plant were to lose its chloroplast, no nuclear element is present that can initiate its *de novo* formation. Chloroplasts, like genes, can undergo mutation and a mutant chloroplast gives rise to a mutant chloroplast, just as mutant genes reproduce mutant genes. This is why variegation is found. It is clear from a study of color mutants and variegation in plants that the cytoplasmic particulate element, the chloroplast, represents a cellular constituent that shows extranuclear inheritance and that the inheritance of a chloroplast requires at the minimum the transmission of a chloroplast itself. Inheritance of a nuclear gene is not sufficient.

We also know, however, that nuclear genes can affect the characteristics of chloroplasts. To explore the relationship of genes and semi-autonomous cytoplasmic particles of this kind, we will discuss an example drawn from another organism. The general characteristics of cytoplasmic inheritance can be clearly demonstrated from the brilliant studies of Tracy Sonneborn and his collaborators. These investigators have studied the genetics of the single cell protozoan, *Paramecium aurelia*. Because certain races of paramecium can kill other races when present together in the same culture medium, certain strains are said to contain a "killer" trait. Strains of paramecium are characterized as either "killer" or "sensitive" types depending on which is the assassin and which is the victim. A detailed genetic analysis (see the book by G. H. Beale in the references) has shown that the killer trait requires a specific element to be present in the cytoplasm. During mating between two paramecia, conditions can be controlled so that some cytoplasm can be transferred between the mating types. If this is done, then the cytoplasmic element associated with the killer trait can be transferred from a killer to a sensitive and transformation of the sensitive animal to a killer occurs. The cytoplasmic element is called "kappa" and the gene known to be required for its replication, K. A killer is genetically KK, the sensitive kk. After mating it can be shown that both sensitives and killers are heterozygous, K/k. In the absence of the cytoplasmic transmission of kappa, sensitives with the genotype K/k, although genetically capable of becoming killers, remain sensitive since they have no kappa, while the killers, with kappa present, remain killers.

These principles were derived in the absence of visual evidence of the presence of kappa. In subsequent experimentation, it has been shown that killer animals do possess a particulate element in their cytoplasm and that this element must be transmitted if the killer trait is to be established. It has further been confirmed that the animals must have a gene (K) in a specific configuration for this element to reproduce and act. In the absence of the transmission of the kappa particle, no nuclear element is present that can initiate the *de novo* formation of this particle. The kappa particle, in turn, requires for its duplication a specific genetic constitution of the animal. The killer trait, therefore, is gene-controlled but not gene-initiated.

Many particulate elements of the cells show cytoplasmic inheritance, and in each case they show the general characteristics found in the inheritance of kappa—i.e., transmission of the trait requires transmission of the particle, but action and/or replication of the particle is controlled by the nuclear constitution of the cell. This mode of inheritance seems to be true for the inheritance of chloroplasts, and from the studies of Boris Ephrussi and Piotr Slonimski, it appears that mitochondria are also inherited this way. It is of interest that recent evidence suggests that chloroplasts and perhaps even mitochondria may have a small amount of DNA. Such DNA might carry the information relative to the structural elements of the organelle. If these observations endure, the inheritance of particulate elements is not in contradiction to our concepts of DNA genetics.

Thus the cytoplasm plays a decisive and intriguing role in determining the alternate and exclusive priorities in gene functioning. The nature of the cytoplasmic control is not understood, but, clearly, the cytoplasm is of importance in inheritance, and its role has many facets, as we shall discuss.

REGULATION

A second problem mentioned at the beginning of this chapter concerns the action of all genetic material. Does all genetic material act alike in determining only the structural characteristics of product enzymes? The evidence is piling up a sound rebuttal. We have known for many years that the expression of some traits is influenced by other genes (modifier genes). It now seems probable that no gene is an island, but is subject to the attention of other genes. It is known that the intensity of pigment formation in plants can be modified so that some strains show more pigment formation than others. These intensity differences are genetically determined but involve many genes, each having a small effect. What do these modifier genes do? They themselves may be genes that control the structure of specific enzymes and that secondarily, even accidentally, affect or modify the trait being measured; or they may be genes that regulate the expression of other genes, perhaps turn them on and off, or change the rate at which they function and thus differ in their action from genes that determine enzyme structure. At present, it seems that both kinds of effects may exist.

First, let us examine a possible mechanism by which a gene can modify a trait through secondary effects. Many cells have an enzyme, beta-galactosidase, that is required for the utilization of the sugar, lactose. Mutation of this gene leads to loss of enzyme activity and loss of ability to use lactose as an energy and carbon source. In order to extract lactose from its surroundings the cell needs a second enzyme. Mutation of the gene controlling this "uptake" enzyme (permease) leads to loss or to decreased efficiency in taking up lactose. So we have a case where mutation of either gene, independent events, leads to inability to use lactose, but for different reasons. If we were not aware of all the enzymes involved, we could well describe the uptake gene as a modifier of beta-galactosidase formation.

Another type of known genetic modification is exemplified by so-called suppressor genes which suppress the expression of other genes. Such genes need not be linked to the gene they affect, and their action may be indirect. As an example, let us assume that the metal zinc inhibits the action of our favorite enzyme, t'ase. A mutation resulting in excessive uptake of zinc would lead to suppression of t'ase activity, and the gene controlling zinc uptake through formation of its own distinct protein would appear to act as a suppressor of t'ase gene action. Like modifier genes, suppressor genes may reflect biochemical interactions rather than diversity in the mechanism of gene action.

A most important mechanism of gene action other than the specification of enzyme structure has emerged from recent observations in bacteria. These studies suggest that some genes serve to regulate by, in effect, turning on and off the action of genes that do specify structure. A model for the mode of action of these "regulator" genes is shown in Fig. 7-1, but to interpret this model we must first describe the operator.

The concept of operator genes stems from the imaginative experiments of the French microbiologist François Jacob and his collaborators. To see how it is envisioned that the operator works, let us designate an active structural gene as enz^+, its mutant allele enz^-, the operator gene as op^+, and its nonfunctional allele op^-. The two genes are closely linked. Two heterozygous diploid combinations are possible and have been experimentally realized:

enz^-	op^+		enz^+	op^+
enz^+	op^-		enz^-	op^-
	trans			cis

Only the cis configuration was found to permit enzyme formation. This in spite of the fact that the structural gene in the trans position was intact in that it could be recovered as an active gene. To work, therefore, the operator requires both linkage and cis configuration. In the diagram we have shown the operator acting on one gene. The operator, however, is not restricted to one gene but it can effect a series of linked genes simultaneously.

We now have the "structural" gene and the "operator" gene, and we require one last unit, the "regulator" gene, to complete the present story. In recent years, inquiry into the metabolism of microorganisms has opened up a field referred to as regulation and it is from the results of these investigations that we have learned of "feed-back inhibition" and "enzyme repression." Feed-back inhibition refers to the phenomenon whereby the end product of a biosynthetic pathway inhibits a key enzyme in the synthesis of this end product. Tryptophan, for example, when made in adequate quantities will prevent a surplus of tryptophan from being made by inhibiting the activity of one of the enzymes in the tryptophan pathway.

Another mechanism of regulation, "enzyme repression," and the one that concerns us here, also functions to alter biosynthesis, but in a different way. Instead of inhibition of an already formed enzyme, as in feed-back inhibition, repression literally turns off the synthesis of enzymes. Space limitations do not permit a full exposé of this field, but recent investigations show that enzyme repression can be triggered by small molecules and it is also clear that genetic components are involved. For example, in *E. coli* a gene, the regulator (R) gene, is known which determines whether the formation of beta-galactosidase is subject to normal control by the operator

Fig. 7-1. Schematic representation of the regulation complex. R = regulator gene. A, B, C are linked genes determining the enzymes of a synthetic pathway. m-RNA$_A$ is the messenger RNA synthesized by the gene A, etc. Production of m-RNA and its arrival in cytoplasm is either permitted or inhibited by the operator gene which, in turn, is turned on or off by the reaction between the regulator product and the small molecule from the cytoplasm.

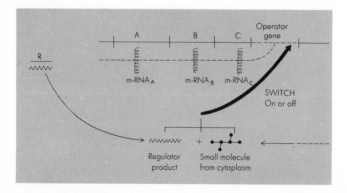

gene. One allele of the R gene permits control by the operator, a second does not. The R gene acts by repressing enzyme formation. Enzyme repression is exerted via the cytoplasm and seems to involve both a specific protein and RNA. A different regulating gene which represses t'ase formation also represses the formation of other enzymes required in the earlier steps of tryptophan biosynthesis, i.e., it can simultaneously regulate the formation of a number of functionally related enzymes. Moreover, this R gene is not linked to the genes whose activity it regulates.

In Fig. 7-1 is presented a summary design that describes in an oversimplified fashion a picture of gene cooperation and regulation. For every step shown there is some evidence but, since none of it is absolutely secure, it is not to be taken as more than schematic. The regulator gene, which may or may not be linked to the genes it controls, directs the synthesis of a product such as RNA. This product specifically interacts with small molecules coming in from the cytoplasm and in so doing determines whether the operator turns its genes off or lets them function. The operator works by interfering with the egress of messenger RNA (m-RNA) to the cytoplasm. If m-RNA flow is inhibited, the structural gene, though intact, is impotent to direct protein formation.

The regulatory product, therefore, is the master molecule in determining enzyme synthesis in the cell. The genome represents the full potential of the cell, and the interaction between regulatory product and molecules from the cytoplasm determines what part of the genome can function. This finally provides an idea, no matter how hazy, of how molecular environment, genetic endowment, and the cell's metabolic activity can influence the potential action of a cell. Molecules entering from the environment or produced by cellular metabolism can reach the nucleus and react with the gene regulator products; this in turn influences the interaction between the regulator product and the operator; and finally the message gets to the structural genes. In Churchill's phrase this "may be the end of the beginning"—the beginning of our understanding of differentiation, development, and the reaction of cells and organisms to their environment. It is a good guess that some of you who read this will have a hand in providing the final solution.

We now again ask, but on a deeper level, do all genes act alike? We have not excluded this possibility. All genes may act alike in that the basic reaction carried out by DNA is the formation of RNA, but not all RNA's are alike in terms of function. As a matter of fact, a little reflection leads to the realization that structural genes have dominated our attention as they are the genes that our present experimental techniques readily detect, but DNA must do more. Present evidence suggests that with the exception of infective RNA, all RNA synthesis is DNA-dependent. DNA makes messenger RNA, this we know. But what about transfer RNA and ribosomal

RNA? These must also be of DNA origin and obviously the composition and types of these RNA's must have a profound influence on the ultimate expression of messenger RNA. Regulator genes may also act via the intermediary of specific RNA's. Simply knowing the diversity of RNA's tells us that DNA does more than transmit structural information. Transmission of structural information is important but not sufficient for a cell that really ticks.

As for the genetics of transfer and ribosomal RNA, for this edition we know very little. Only one mutation has been detected to date which affects the composition of the transfer RNA. In fact we still don't know how many transfer RNA's a cell has. Our one gene is a bare beginning, but it indicates a genetic control of transfer RNA.

We know still less about the origin or synthesis of ribosomal RNA. There is evidence that ribosomal RNA is in the nucleolus, but the nucleolus in turn has a chromosomal organizing region. Whether there is but one type of ribosome or many we don't know, but what we can say is that it is of DNA origin and subject to mutation.

The fact that transfer and ribosomal RNA are of genetic origin brings in its wake some curious possibilities. For instance, we know a good deal about the code of messenger RNA and we assume that the transfer RNA has a complementary triplet code, but this isn't necessary. Transfer RNA could use a doublet code and it could still interact with a triplet messenger code. The point we wish to make is simply that expression of genetic information requires the interactions of messenger, transfer, and ribosomal RNA. Each may have a unique specificity and the extraordinary specificity we observe is in reality an interaction of three different RNA's, each of separate genetic origin and each possessing unique specificities.

Once again, we must emphasize that although the studies of the past ten years in the field of molecular genetics have been extraordinarily rewarding, the information obtained does not provide the solution for many problems in biology, particularly those concerning differentiation and development. It may well be that genetic elements exist whose actions have not yet been described; we don't know. At the minimum, development requires a full understanding of genetic regulation and of the interactions between nuclear and cytoplasmic elements. Genetics and development constitute a field of tremendous experimental possibilities, a field in which many new biological laws are likely to be discovered in the future.

Genetic
Mechanisms

In the preceding chapter, our emphasis has been on the gene itself, its structure and its function. We now turn to a consideration of the mechanics of gene transmission and the laws of heredity. We introduce this subject by again invoking the cooperation of our favorite haploid organism, *Neurospora crassa*. In neurospora, the vegetative transmission of traits takes place exclusively by mitotic nuclear divisions. As the organism grows, nuclei divide, and, as the number of nuclei increases, the mycelial mass expands. As we already know, in a well-behaved mitotic division, without chromosomal imbalance or gene mutation, two daughter nuclei are formed identical to the parent nucleus.

Since the growth phase of neurospora is haploid, each nucleus contains only one set of chromosomes. Each chromosome differs from all the others in the genes it carries and, therefore, in the set of traits it determines. When gene mutation occurs in any haploid nucleus, an immediate alteration of gene function in that nucleus must take place.

MUTATION AND SELECTION

Whether or not such a nucleus survives depends on many factors. To illustrate, assume that we have two neurospora conidiospores, each containing one identical nucleus. They differ, however, in a single respect. In one of the spores, a mutation of the tryptophan synthetase gene has occurred, inducing a nutritional requirement for tryptophan. If we now place both spores in an environment that normally permits germination but that contains no available source of tryptophan, what happens is easily predictable. The normal spore will germinate to give rise to a new plant; the mutant spore will not germinate, and its nucleus, with its mutant gene, will be lost.

Consider another type of mutational event. Assume that a single mutation occurs in one spore to confer upon it the capacity to germinate and to grow more rapidly in a relatively arid environment. If the two spores are placed in a normal environment but with a climate in which the moisture content is reduced, the mutant spore will germinate and the plant will grow faster than the normal spore. The mutant will extract and utilize all the resources of the environment at a more rapid rate than will the normal plant. In sum, the mutant will outgrow the normal plant and displace it from the new environment. In this case, the normal nucleus with the normal gene will be eliminated.

In the above examples, one or the other of the parent and mutant nuclei was eliminated. These examples, although somewhat simplified, illustrate a principle that has been validated by countless observations and experiments. Mutation and competition in the original or in a new environment represent powerful forces in the creation and survival of new, fitter genotypes in a haploid organism. The transcendent importance of the sexual phase in neurospora, and in all organisms, lies in the creation of endless new gene combinations. The continuing generation of new gene combinations, whether by mutation or sexual recombination, or both, represents the ultimate in life's resources in facing and responding to a challenging environment. It gives us an indication of how life, on this planet alone, has been able to accommodate itself to such an incredible variety of conditions. And it represents the reserve potential for survival under drastically altered conditions.

BACTERIAL RECOMBINATION

We have shown how new gene combinations form by genetic recombination in an organism that enjoys meiosis. The inquiring mind, however, might wonder about those organisms that spend their entire life as haploids, organisms such as bacteria and bacterial viruses. Can organisms such as these also undergo genetic recombination or must they evolve new genotypes by mutation alone? Experimental work of the past decade has shown that bacterial genes are linked and must therefore be organized in structures that are similar to the structure of a chromosome. This fact has been deduced from genetic experiments, although it has not yet been confirmed by cytological examination.

Escherichia coli and *Salmonella typhimurium* are examples of two bacterial species which characteristically contain one linkage group and are haploid. Vegetative reproduction takes place through duplication of the bacterial chromosome and the transmission of an identical chromosome to each daughter cell. We suspect that this event is similar to mitosis, although this conclusion lacks cytological confirmation.

As we will discuss in a moment, mutation plays a major role in the evolution of new genotypes in these organisms. However, bacteria are also capable of recombining the genetic traits of two separate cells. In bacterial transformation (see Chapter Two), certain bacteria can take up purified DNA from a genotypically different cell, and the new information carried by the DNA can be incorporated by some mechanism of recombination into the recipient cell's own genetic apparatus and transmitted to its progeny. In transformation, therefore, a cell is formed that can be called, with some charity, heterozygous, as it contains simultaneously, even though fleetingly, two alleles of a given gene.

Regardless of the recombination mechanism involved, however, recombination such as that described for bacterial transformation clearly does occur, and the formation of partial heterozygotes is characteristic of genetic recombination in the bacterial world in general. It is the *manner* in which donor DNA is acquired by a recipient cell that varies. In transformation, the donor cell DNA is picked up from the medium. Another mechanism for recombination, bacterial conjugation, was discovered by Joshua Lederberg in certain strains in *E. coli*. In these strains, two cells can fuse and form a cytoplasmic bridge. The chromosome of the donor cell is then injected into the recipient cell. Bacterial conjugation requires a mating-type system, since cells of only one kind (Hfr) can serve as donors, while other cells (F⁻) can serve as recipients. The difference between (Hfr) and (F⁻) cells lies in the possession by the (Hfr) of an episome. Episomes are remarkable cellular elements that enjoy both a chromosomal and cytoplasmic existence, and are exemplified by temperate viruses, as you can read about in the next paragraph. Surprisingly enough, even under conditions in which a cytoplasmic bridge is formed between the two cells, the entire genome of the donor cell is seldom injected. Again, the recipient cell is only partially diploid. After conjugation, the cells separate, and recombination takes place between the genes injected by the donor cell and the genes present in the recipient chromosome. Daughter cells bearing new combinations of the two parent cells are formed, and the new combinations are transmitted to their progeny.

Still a third mechanism for recombination, bacterial transduction, exists in bacteria. Bacterial transduction differs from the other mechanisms in that the agent for transmitting a portion of the DNA from one cell to another is a virus. It may be somewhat disquieting to think that a virus, the very model of a cell parasite, can transfer its host's DNA from one cell to another. It does occur, however, even if unexpected. The story is briefly this. Although many bacterial viruses appear to infect and invariably destroy the specific bacterium they parasitize, other viruses behave differently. These can and do infect and destroy, but, under some unexplained circumstances, they infect cells and then turn benign. The viral DNA becomes attached to the chromosome of the host cell and multiplies when the host chromosome multiplies. If a virus has this characteristic, it is known as a temperate virus. The condition whereby the viral DNA sits quietly on the host chromosome is called *lysogeny*.

From time to time, the virus spontaneously erupts, takes up a cytoplasmic existence, reassumes its malignant form, multiplies, destroys its host cell, and pours out ready to infect another host cell, prepared either to destroy it or to lysogenize. When the DNA of a temperate virus in a lysogenic condition is activated, it forms a complete virus. It can then lysogenize a new host and carry a piece of its old host's chromosome into the new host. Moreover, the genetic information of this chromosomal fragment can become incorporated into the new host.

Bacterial recombination, therefore, characteristically takes place through a number of different mechanisms, but only partial diploid cells are formed. The absence of equal genetic contribution and the absence of a true reduction division distinguish bacterial recombination from that found in higher organisms.

One of the many enigmas that surround these various forms of bacterial recombination is the role they play in nature. Do they represent laboratory oddities forced upon nature by the investigator or are they crucial to the survival of cells? And of equal interest: Are these exclusively bacterial phenomena? We know that human cells are susceptible to certain types of virus and that they can also harbor viruses for years, a condition analogous to lysogeny. Do human viruses carry DNA from cell to cell and, if they do, why? These puzzles are not meant to glorify the status of the question mark, but to give you some idea of the things that intrigue biologists at the moment, and of the things they are trying to learn. Therefore, although we know that bacterial recombination can occur, its role in the creation of new bacterial types has still to be assessed.

At the present time, it would appear that mutation and selection are the major forces involved in the evolution of new bacterial cell types. We know that mutation occurs spontaneously and in all genes that have been studied, although the spontaneous rate of mutation among genes appears to vary. In general, bacteria are single haploid cells. Mutations are rapidly expressed. Moreover, haploid organisms multiply very rapidly as compared with the higher diploid forms. In the laboratory, under suitable conditions, the bacterium *E. coli* will divide about every half hour. At the end of 15 hours, one cell will give rise to 10^9, or about one billion cells. This means that mutant populations can easily be expressed, and selection of new genotypes by the environment can easily occur. We have illustrated the principle of mutation and selection earlier in the chapter and will now discuss one additional example which has come to our attention rather forcefully in recent years.

Escherichia coli is ordinarily killed by the antibiotic substance streptomycin. If streptomycin is added to a culture of sensitive cells, the cells will be killed. However, *E. coli* contains a gene which can undergo mutation and confer resistance to streptomycin upon cells that carry it. Mutation to streptomycin-resistance occurs spontaneously, about once in every 10^6 to 10^9 cell divisions. In the absence of streptomycin in the environment, a mutation from streptomycin-sensitivity to streptomycin-resistance will go unnoticed by us, and apparently also by nature, since the streptomycin-resistant genotype does not accumulate to any noticeable extent in normal populations of *E. coli*. In the presence of streptomycin, however, only those cells with the resistant gene can survive and grow, and the streptomycin-sensitive population is eradicated. In a world filled with streptomycin, the mutant gene becomes the normal gene.

This is an important and contemporary problem. Microorganisms have a remarkable genetic flexibility, the potential to mutate and meet the challenge of as yet unknown environments in order to survive. In a similar way, as you must know, the flies have conquered DDT, and now man is considering whether he can conquer pesticides, and the ionizing radiations, the weightlessness, and the new challenges of space.

In the analysis of gene action and the brief discussion of genetic mechanisms, we have concentrated almost exclusively on haploid organisms or those with but a passing diploid phase. The reason is obvious. Genetics is young, life is complicated, and we need the simplest system we can find to develop our still naive idea. But, curiosity will not be denied, and we want to take a look at higher forms of life, a category in which we are not reluctant to place ourselves. To do this, we now turn to the elements of diploid genetics.

DIPLOIDY

The laws of genetics were originally formulated from the study of higher plants and animals. These organisms differ from fungi and bacteria in that they are diploid and invariably possess some form of sexual reproduction. Typically, a new individual arises by fusion of two nonidentical gametic cells, the egg and the sperm cell. Fusion results in the formation of a diploid cell, which, by division, growth, and differentiation, gives rise to a new adult. Reproduction, therefore, requires the formation of haploid gametes from the diploid cells of a mature organism. The formation of progeny consists of two meiotic divisions, one in the male and one in the female cell lines. The fact that in the diploid type of life cycle a new individual arises by fusion of two nonidentical gametic cells suggests that recombination plays a major role in the formation of new types.

The formal genetics of diploid organisms differs in certain respects from the genetics of haploid organisms. In diploid genetics, the observable characteristics of the organism (the phenotype) derive from the existence and expression in each cell of a duplicate set of genes, each pair of which may be identical (homozygous) or different (heterozygous).

DOMINANCE

To illustrate this problem, let us choose a microorganism, a yeast, whose vegetative growth phase is diploid.

Diploid yeast cells divide by mitosis and can also undergo meiosis to form haploid cells. The fusion of two haploid cells results in the re-establishment of the diploid line. Two haploid strains, both capable of forming tryptophan synthetase (t'ase$^+$), upon fusion, form a diploid cell carrying two t'ase$^+$ genes. Two haploid strains incapable of forming the enzyme (t'ase$^-$) fuse to give a diploid cell line with a nutritional requirement for tryptophan. In a fusion between a t'ase$^-$ strain and a t'ase$^+$ strain, the diploid line now carries both alleles; on one chromosome is a gene that can direct the formation of the enzyme, whereas on the homologous chromosome is a mutant allele that cannot. The phenotype of the diploid cell line, which is heterozygous for the tryptophan synthetase gene, is tryptophan-independent. The cells synthesize the enzyme and can grow in the absence of added tryptophan. The t'ase$^+$ gene is said to be dominant over the t'ase$^-$ gene.

In general, the ability to synthesize an active enzyme qualifies a gene as dominant. The presence of a dominant gene in the diploid cell also usually permits the formation of a specific enzyme at a rate sufficient to allow growth, although not necessarily at the maximum possible rate. This example illustrates the problems, however, that can arise as a consequence of heterozygosity. For instance, a mutation in the t'ase$^+$ gene can result in the formation of a mutant enzyme that might in some way inhibit the normal enzyme. This would lead to a requirement for tryptophan in a heterozygote containing a normal t'ase$^+$ allele. In the absence of other knowledge, we would then conclude that the mutant gene was dominant to the normal allele.

The relationship of dominance and its converse, recessiveness, is a consequence of the fact that in diploid cells, each one of the alleles exerts its own action, and the phenotype must reflect the result of their interaction. If one of the alleles can function and the second allele cannot, and is neutral, the former would appear dominant. If the second allele inhibits the action of the first, the second would appear dominant. If one allele can direct the formation of a product but in insufficient amount, the heterozygote appears intermediate, and dominance is indeterminate.

The fact that a vegetative cell carries duplicate genes confers genetic stability on a cell, and this may be one of the reasons that the majority of higher plants and animals are diploid. In haploid organisms, such as neurospora and bacteria, mutation can lead to the rapid phenotypic expression of the mutant allele. In a diploid organism, however, expression of a recessive mutation requires that the mutation occur at least twice, either in both members of a diploid pair or singly in two diploid genomes followed by recombination. If the probability of mutation of one member of a gene pair is 1 in 10^7 cell divisions, the probability that both members of the gene pair will mutate in the same cell at the same time is a product of this probability, or 1 in 10^{14} cell divisions. This is a rare event. In addition, if the mutation occurs separately, the mutant alleles must find each other or they may be lost in the meiotic shuffle.

Thus, many mutations can occur in a diploid organism and pass undetected during the vegetative propagation of this organism. The strain can simply grow and accumulate recessive mutations. Occasionally, a double recessive will be formed, and the phenotypic expression of the gene can then be observed. If it is valuable, we treat the organism and the parent lines with great deference. If it is troublesome, we try to correct or eliminate it. These are common practices in horticulture and animal husbandry.

Although diploidy confers great stability on vegetative reproduction, sex combined with diploidy provides the opportunity for a species to try out many new gene combinations. To see how this works, let us consider the transmission of traits in a diploid organism in which the two diploid cell lines both undergo meiosis and form haploid gametes.

THREE:ONE

Assume that two cell lines differ in a particular trait—enzyme formation— and that a dominance relation holds. Let E represent the (dominant) gene which directs enzyme formation (enz$^+$) in the phenotype, and let e represent the (recessive) inert allele. Assume that one organism has the genotype EE and the other ee. A cross between the two will give the results outlined in Figure 8-1A, where P stands for parent, G_m for gamete produced by meiosis, and F_1 and F_2 for the first and second filial generations. In this case each parent produces but one type of gamete, and all the F_1 progeny are alike, Ee (enz$^+$).

A cross of two F_1 organisms is diagrammed in Fig. 8-1B. Here, each

F_1 gives rise to two gametes. These are usually produced in equal number. Random fertilization means that fusion of any gamete from one cell line by any gamete of any other cell line is equally probable. Random fertilization of F_1 gametes produces three F_2 genotypes in the ratio shown in the figure. Since both *EE* and *Ee* are (enz$^+$), there are only *two* phenotypes in 3:1 ratio. In the F_2, note the reappearance of the two parent types, *EE* and *ee,* and the persistence of the recombinant type, *Ee*. In diploid genetics, a 3:1 phenotypic ratio in the F_2 is a clear indication of a single-gene difference.

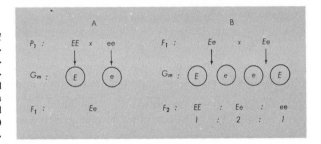

Fig. 8-1. The inheritance of a single gene. P₁ = Parental strains, Gₘ = Gametes, F₁ = First filial generation, F₂ = Second filial generation. (A) Mating of a homozygous dominant and homozygous recessive. (B) Mating of F₁ heterozygotes.

NINE:THREE:THREE:ONE

Let us now analyze a cross involving two gene differences. Since in this book we love enzymes, let the traits controlled by the genes be: T = the presence of t'ase (t'ase$^+$), t = no t'ase (t'ase$^-$), H = the presence of histidase (h'ase$^+$), h = no h'ase (h'ase$^-$). Assume that the T and H genes are on separate chromosomes.

The results of a cross between the parental types, *TTHH* × *tthh,* are given in Fig. 8-2, as is the result of the F_1 cross. By now, you have all the information necessary to calculate and confirm the ratios and the genotypes given in Fig. 8-2. Moreover, you can backcross the F_1 to either parent and predict the results. You can check your answers by referring to any of the standard textbooks given in the bibliography. The appearance of a 9:3:3:1 ratio (with dominance in each gene) clearly indicates the segregation of two unlinked genes.

LINKAGE

For our third example, we will analyze the transmission of two genes linked together on the same chromosome. Chromosome assortment during meiosis would *by itself* not result in the separation (segregation) of linked genes. In the absence of crossover, all the progeny would be phenotypically and genotypically like their parents. Assume that the tryptophan and histidine genes are linked on the same chromosome. A cross of the homozygous dominant $(T,H/T,H)$ by the homozygous recessive $(t,h/t,h)$ will give heterozygous F_1 progeny $(T,H/t,h)$ (Fig. 8-3). Meiosis in the heterozygous progeny will normally result in the formation of only two types of gametes, (T,H) and (t,h). Thus, if two heterozygous F_1 progeny are crossed, we will find in the F_2 only strains that are $(T,H/t,h)$, $(T,H/T,H)$, and $(t,h/t,h)$, all phenotypically like one or the other parent. No segregation of the parental genes on the chromosome has occurred.

However, as we discussed in Chapter Three, during meiosis the intimate pairing between homologous chromosomes permits the occurrence of segmental exchanges, the phenomenon of crossing over. The consequences of crossing over, at an assumed frequency of one in five, are given in full in Fig. 8-3.

By examining the figure, you will notice that all the F_1 progeny are dominant for both genes, as is true of the F_1 progeny that arise when the genes are in separate chromosomes. The difference, however, is that both dominants are still on *one* chromosome. In the F_1 cross, recombination of the dominant and recessive genes takes place through crossing over. (Have you noticed that crossing over in the parental chromosomes has no genetic effect?) The types of gametes formed, their frequencies, and the full genotypic and phenotypic consequences of random fertilization are re-

Fig. 8-2. The inheritance of two genes on separate chromosomes.

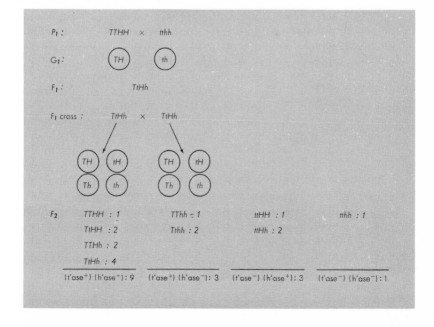

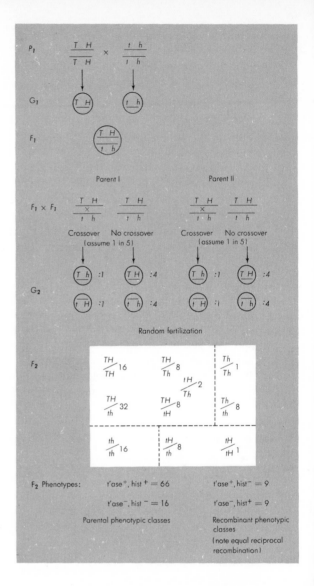

Fig. 8-3. The inheritance of two linked genes assuming 20 per cent crossover.

corded in Fig. 8-3. Note that nine different genotypes are produced in the F_2 by the breakage-fusion recombination. Genes, even when linked, do segregate. Thus the final phenotypic ratio with 20 per cent (1 in 5) crossing over is 66:9:9:16, far different from the 9:3:3:1 ratio expected for unlinked genes. Study the figure carefully and devote some thought to it. Set up the following problem:

$$\left(\frac{T\ H}{t\ h}\right) \times \left(\frac{t\ h}{T\ H}\right)$$

Assume a 20 per cent crossover frequency and solve it by writing every resulting genotype, phenotype, and the expected ratios. (For our illustration, a crossover frequency of 20 per cent was assumed, but remember that the frequency is a function of the distance between the genes on the chromosome.) You will be in a position to understand some of the fascinating work that has been done, and is still being done, in the genetics of higher organisms, in breeding and selection, in insect population control, and in the effects of radiation.

SEX DETERMINATION

Before we go on to a study of human genetics, we must discuss two additional subjects: the genetic determination of sex and a curious event, chromosomal nondisjunction, first observed in studies on sex determination. In microorganisms, sexual reproduction arises from a mating-type system. In general, there are no important morphological differences between the mating types. They are characterized by the fact that strains within a mating type are mutually infertile. Strains from different mating types will crossbreed to form a partial or complete zygote. This compatibility difference is inherited as a single-gene difference. In true sexuality, clear functional and structural differences exist. There are two gametic cells of different function, the egg cell and the sperm cell, with elaborate functional differences between the adults of the two sexes. In humans, many developmental differences are known to exist between the sexes, and these differences appear to reflect the action of many different genes, i.e., sex determination is multigenic. This is borne out by the fact that sex determination involves a chromosomal difference with its full complement of genes.

The chromosomal basis of sex determination can be seen clearly in the fruit fly, Drosophila. Cells of the female have four homologous pairs of chromosomes, while cells of the male contain three homologous pairs (autosomes) and one unlike pair. The unlike pair consists of one chromosome that is homologous to one in the female cells and another that is distinctly dissimilar and found only in male cells. The chromosome found in both cells is called the X chromosome, and the unlike chromosome characteristic of male cells is called the Y chromosome. Females can transmit only X chromosomes, while males transmit either X or Y, and in equal frequency. Clearly, then, there is an equal probability that a newly fertilized zygote will contain either two X chromosomes, a new female, or one X and one Y, a new male, thus providing for equality in numbers of both sexes.

The chromosomal basis of sex determination in man is superficially similar to that in Drosophila. The female has two X chromosomes in her cells, the male an X and Y. This chromosomal distribution, however, is not universal. In some organisms—the domestic fowl and some moths—the male has the equivalent of two X's and the female the X and Y. In still others, the female normally has two sex chromosomes and the male only one—there is no "Y" chromosome and there are other variations.

The role played by sex chromosomes was clarified by a study of non-disjunction by Calvin Bridges. He found that on rare occasions, during meiosis, the X chromosomes paired, as expected, but then *did not separate,* did not disjoin. They both went into one of the succeeding daughter cells. The other cell did not receive an X. These eggs, XX and O instead of the normal X, were then fertilized by either X or Y from the male to give XXX, XXY, OX, and OY. In Drosophila, XXX developed into an infertile female (called a superfemale—unfortunately, a most inappropriate term); XXY was a fertile female, OX a sterile male, while OY was lethal, the egg did not hatch.

It is now known that nondisjunction can occur in many organisms, including man. Of particular interest is the fact that it apparently can involve any chromosome. The revelations provided by studies of nondisjunction in man in terms of both sex determination and the development of physical and mental abnormalities will be discussed in the next chapter.

Genes and Man

CHAPTER NINE

History tells us that not very long ago man thought of his remote planet as the center of all things with the rest of the universe revolving respectfully around it. Scientific inquiry has demolished this small conceit but not man's ego. It is no wonder that he has made human genetics a subject of special interest. And it is in this part of genetics that he needs ingenuity, for, in obedience to cultural requirements, he must deny himself the free use of a valuable experimental tool, controlled mating.

Direct study of the genetic basis of human traits has depended largely on refined statistical analysis of genealogical lines and the compilation of vital statistics. Fortunately, all the evidence accumulated by these methods confirms the proposition that the transmission and function of human genes corresponds to that of other biparental diploid organisms. Let us begin with an examination of some very recent data which has derived from the introduction of new and promising methods.

THE CHROMOSOMAL BASIS
OF GENE TRANSMISSION

Surprisingly, the number of chromosomes characteristic of the human species has long been a subject of controversy. Identification is difficult because many of the chromosomes are cytologically alike, while others are small and difficult to detect. Recently, human cells have been cultured in the laboratory by techniques (tissue culture) similar to those used with microorganisms, and, under these conditions, they provide excellent material for cytological investigation. By special methods, cell division can be arrested and the chromosomes maintained at the most favorable point for observation—on the equatorial plane just before separation. These studies reveal a basic chromosome number in man of 46, consisting of 23 pairs. Cells derived from females show 23 homologous pairs, while cells from males show 22 pairs and a twenty-third pair consisting of an X-chromosome and a shortened Y (Fig. 9-1). An immediate result of the development of tissue culture methods was the discovery in man of nondisjunction.

SOME CONGENITAL DISEASES
AND NONDISJUNCTION

It has recently been found that Mongolism, a serious neurological disorder accompanied by a characteristic Mongoloid appearance and by mental

Fig. 9-1. Photomicrograph of the chromosome of a human male. (Courtesy Dr. J. H. Tjio.)

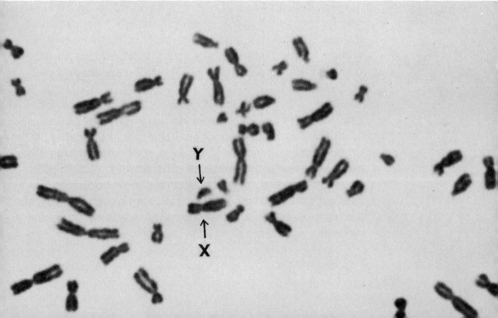

retardation, is associated in many cases with a chromosome count of 47. The rare child with 47 chromosomes is generally born to parents with normal counts of 46. The most reasonable explanation for this aberration is nondisjunction. An egg cell in which one of the chromosome pairs does not separate during meiosis would, upon fertilization, form a zygote with 47 chromosomes, 22 normal pairs and 1 triplet. In Mongolism, the triplet normally occurs in chromosome No. 21, and, for some unexplained reason, the additional chromosome alters development and results in the described neurological disorders.

Nondisjunction of other chromosomes, including the sex chromosomes, has been found and is associated with other disorders. Individuals are known whose cells carry two X-chromosomes and one Y. They suffer from a neurological defect known as Kleinfelter's Syndrome and show physiological and psychological abnormalities relating to sex determination and expression.

Nondisjunction of the sex chromosomes implies an interesting consequence. If a pair of chromosomes does not separate during meiosis, then one of the products of meiosis gets both chromosomes and another should get none of the chromosomes. When fertilized, the latter would have a chromosome number of 45 and be haploid for the missing chromosome. This has been verified. Individuals have been found whose cells contain an X but no Y. Phenotypically, they are undeveloped females with no ovarian tissues (Turner's Syndrome). In addition to providing evidence for nondisjunction, the discovery of XO offers some interesting revelations into sex determination in man.

In Table 9-1 are listed some known genotypes relating to the sex chromosomes and the corresponding phenotypic expression in both man and Drosophila.

Table 9-1

THE PHENOTYPIC EXPRESSION OF VARIOUS GENOTYPES
IN MAN AND DROSOPHILA

Genotype ⟶	X	XX	XY	XXY	XXXXXY
Human being	UF	F	M	M	M
Fruit fly	M	F	M	F	

F = Female   M = Male   UF = Undeveloped Female

Notice that in man, X and XX are female, whereas the possession of only one Y is sufficient to confer the male phenotype. In Drosophila, one X, with or without Y, is male, while XXY is female. From this and additional evidence, it appears that Y in Drosophila carries few, if any, genes and none required for male development. In human beings, the genes carried on Y appear to exert a powerful influence in sex determination, as is attested by a male with the bizarre chromosomal constitution, 5X,Y. In general, it can be concluded that the possession of an extra chromosome or the omission of one of the chromosomes leads to developmental abnormalities. The difficulty appears to be one of imbalance. In the haploid, diploid, or polyploid state, in which full complements of chromosomes are present, gene products are apparently formed in varying amounts, but in a balanced condition. An additional chromosome seems to disorganize this situation. This is another indication that genes do not act independently of one another in determining the phenotype of an organism, and that the balance of structure and regulatory genes may well be critical.

Our discussion has touched on but a limited part of the evidence available from intensive researches now under way into the chromosomal basis of human heredity. In the future, many more instances of the association of a specific disease pattern with chromosomal imbalance will probably be found. We know no treatment for this type of disease. Perhaps one day it will be possible to prevent nondisjunction or to rectify the physiological consequences of nondisjunction once it has occurred. These are not idle hopes. Other human disorders with a different genetic basis are known. By making use of the facts and principles discussed in the preceding chapters, we have been able to establish a rational course of therapy in some of these cases. As an illustration, we shall discuss a disease which is called "phenylketonuria."

PHENYLKETONURIA

In the 1930's, it was observed that unusual amounts of a compound, phenylpyruvic acid, were excreted in the urine of certain patients afflicted with severe mental disorders, usually idiocy or imbecility. The disease was called phenylketonuria and appeared to affect members of the same family. A more detailed analysis of the pattern of family distribution turned up a result unexpected at that time. The inheritance of phenylketonuria could be explained on the basis of a single-gene difference. It was found that phenylketonuria occurred in brothers and sisters of affected persons, but rarely their parents or in more distant relatives such as uncles, aunts, and cousins. The disorder was unusually frequent in consanguineous marriages, i.e., those between cousin-cousin, uncle-niece, aunt-nephew. (You realize that consanguineous mates have a common ancestor.) Furthermore, even in affected families, phenylketonurics were in the minority.

The observations all pointed to the association of this disease with a recessive allele of a single normal gene. Three genotypes could then exist in the population, the homozygous dominant with a normal phenotype, the heterozygote, also normal, and the homozygous recessive, the abnormal phenotype. The homozygous recessive condition accounts for the increased frequency of the disease in consanguineous marriages, since the probability is increased of a mating between two heterozygous individuals to give the required homozygous recessive. Once the genetics of the transmission of the disease had been solved, the rest became a matter of biochemical detection.

The biochemical basis of phenylketonuria lies in the metabolism of the amino acid phenylalanine (Fig. 9-2). Part of the phenylalanine we ingest

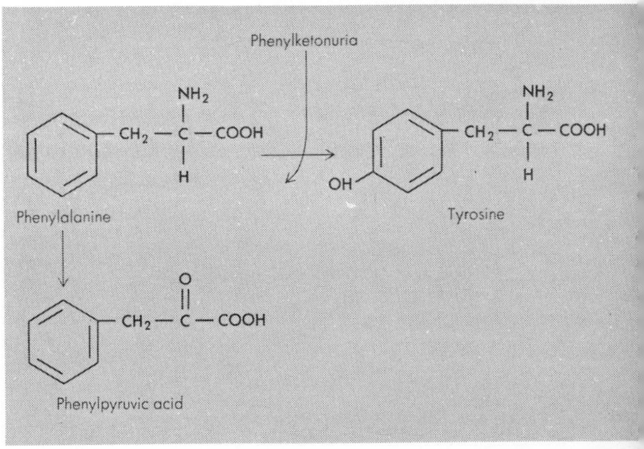

Fig. 9-2. The biochemical basis of phenylketonuria.

in our diets is normally oxidized to another amino acid, tyrosine. If, for some reason, the oxidation is blocked, phenylalanine is alternatively oxidized to phenylpyruvic acid, which is then excreted. Combine the genetic evidence (a single gene) with the loss of a single reaction and our suspicion falls, correctly, on an enzyme. The recessive allele cannot direct the formation of the enzyme that catalyzes the conversion of phenylalanine to tyrosine. But how can this single loss lead to mental impairment?

Human beings are unable to synthesize many amino acids essential for protein formation—they must obtain them in their diet. Phenylalanine is one of these required amino acids, but tyrosine is not, since a sufficient amount of tyrosine is ordinarily formed by the oxidation of dietary phenylalanine. The normal diet, however, provides both phenylalanine and an adequate amount of tyrosine, and thus the genetic loss of ability to convert phenylalanine to tyrosine does not impair growth. By itself, the loss of the enzyme should be harmless. However, since phenylalanine cannot be converted to tyrosine, an alternative mechanism takes over and causes the production of abnormal amounts of phenylpyruvic acid. The accumulation of phenylpyruvic acid, a secondary consequence of enzyme loss, appears to be responsible for the neurological damage.

The course of therapy evolved for this disease makes use of all the information outlined above. If this disorder involves just the one enzymatic defect, it should be possible to rear such an individual without attendant mental impairment by maintaining dietary levels of phenylalanine just sufficient for protein synthesis, but low enough to keep the phenylpyruvic acid at a minimal and, hopefully, nontoxic level.

This treatment has only recently been applied, and the data gathered so far are encouraging. Phenylketonuric children reared from birth on a carefully prepared diet containing subsistance levels of phenylalanine appear to grow normally, both physically and neurologically. Observations on their progress are still continuing, and there is reasonable hope that the disease may have been circumvented.

Phenylketonuria offers a valuable lesson when we are considering the effects of gene mutation. A genetic change that affects a specific biochemical reaction may indirectly bring about more extensive biochemical changes than the loss or inhibition of one reaction. If a cell loses its normal ability to carry out a reaction, say $A \longrightarrow B$, compound A may accumulate or may be metabolized to other compounds. Either event can have a profound metabolic effect and must be taken into account when assessing the over-all phenotypic effect of a single-gene mutation. Phenylketonuria also makes it clear that if a human disorder can be shown to correspond to a single-gene difference, there is a strong probability that the difference between a normal and an abnormal individual is basically a difference in a single enzyme. If we can recognize the reaction, we may be able to bypass the genetic defect.

The inheritance of galactosemia is another case in point. Galactosemia, a disease first noted in children, is inherited as a single-gene difference and involves the loss of a single known enzyme. As in the case of phenylketonuria, we find that the loss of a single biochemical reaction indirectly leads to a number of associated abnormalities, including mental retardation. Knowledge of the biochemical basis of the defect again permits rational treatment. The interested student can readily find the details in a reference given at the end of the book.

From these examples, we can see that the basis of gene action in man, control of biochemical reactions through the mediation of enzyme formation, is the same as that in microorganisms and other subdivisions of the plant and animal kingdoms.

SICKLE-CELL ANEMIA

In the cases cited above of inherited disease in man, we drew on information gathered from other sources, notably microorganisms, to clarify the nature of the disease and to prescribe treatment. In turn, a series of brilliant studies on another human disease, sickle-cell anemia, has significantly increased our knowledge of gene action in all organisms.

In 1949, Linus Pauling and his colleagues discovered that the formation of an abnormal hemoglobin appeared to be the biochemical basis for the inherited trait, sickle-cell anemia. As you know, hemoglobin is the major protein of red blood cells. It combines with oxygen in the lungs and carries it to all parts of the body. In general, a decrease in the amount of atmospheric oxygen has no effect on the red blood cells of normal individuals. However, it was noted as early as 1910 that with a decrease the red cells of some people become elongated and take on an odd sickle shape. Ordinarily this does not affect the health of the individual, but in some cases red cell sickling is associated with severe hemolytic anemia. A study of the family distribution of red cell sickling showed it to be inherited as a single recessive gene. It was also possible to distinguish heterozygous individuals from homozygous recessives, since heterozygotes show sickling but no symptoms while homozygous recessives show sickling and suffer from hemolytic anemia.

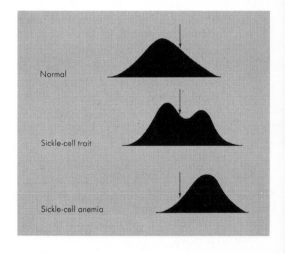

Fig. 9-3. Electrophoretic behavior of various human hemoglobins (after Pauling, et al.). Normal—homozygous dominant; sickle-cell trait—heterozygote; sickle-cell anemia—homozygous recessive. Arrow indicates reference point of origin.

Pauling and his collaborators isolated the hemoglobin from normal and sickled cells—the latter from homozygous recessives—and discovered that the two hemoglobins differed in the electric charge on the surface of the molecules (Fig. 9-3). The hemoglobin of the heterozygote was then isolated and found to be composed of about equal amounts of both normal and abnormal hemoglobin (a nice demonstration that both alleles were functioning independently of each other). With the help of fingerprinting and amino acid sequence analysis as described in Chapter Four, Vernon Ingram showed that the two hemoglobins differed in one amino acid out of about five hundred. One glutamic acid residue in normal hemoglobin,

$$H_2C\!-\!COO^-$$
$$H_2C \qquad O$$
$$-NH\!-\!CH\!-\!C\!\!-$$

is replaced by lysine,

$$H_2C\!-\!CH_2\!-\!NH_3{}^+$$
$$H_2C$$
$$H_2C \qquad O$$
$$-NH\!-\!CH\!-\!C\!\!-$$

in precisely the same place in the protein to give sickle-cell hemoglobin. The amino acids are written with the charges they would have in the red cell, and explain the observed charge difference.

The substitution of glutamic acid by lysine is the only difference between the two hemoglobins, but this replacement has a profound effect on the physical characteristics of the formed protein. It results in the alteration of electrical charge as well as in a difference in biological activity, namely the transport of oxygen. Study of still other human hemoglobins has brought to light many other single amino acid differences, telling us that in terms of gene action human beings are as well behaved as *E. coli* and neurospora.

Sickle trait reveals a second, unrelated bit of human biology. One might well ask why, if sickle hemoglobin is inefficient in transporting oxygen, has this allele remained in human populations? It turns out that this allele affords a selective advantage to its possessors in a region infested with malaria. Individuals carrying the sickle trait are not as susceptible to malaria as those carrying the normal allele. In malaria-ridden regions, the replacement of glutamic acid by lysine thus confers indefinite selective advantage as this replacement indirectly enhances the likelihood of surviving to the age of reproduction. Why this should be so is a fascinating story for the future!

MULTIPLE ALLELES

The most striking case of multiple allelism in man occurs in the genes that determine his blood types. Blood characteristics differ from individual to individual, and one can observe this by a simple test. Mix a sample of red blood cells from one individual with serum from another (it would probably be advantageous at this point to reread the section on immunization in Chapter Four). A difference in blood type between the individuals will be revealed by the clumping of the red cells in the serum. The red cells will form large aggregates and settle to the bottom of the tube. This phenomenon is called agglutination and indicates incompatibility between the bloods. Agglutination can be seen in the test tube, but will also occur in the blood stream; i.e., if blood is injected from one individual into a second and the blood is incompatible, agglutination of the red blood cells of the donor will occur. Since agglutination can be fatal, careful tests must be made to determine what blood may be safely used in transfusion.

Agglutination requires two components, a component in the serum and a component on the cells. The elements in the serum that are concerned with specific agglutinating activity are called antibodies and were described in Chapter Four. The specific properties of the cells that enable them to react with antibodies are antigens, and again these were discussed in the earlier chapter. When the cells and the serum of two individuals, A and B, are mixed in all possible combinations, the cells of A, when put into the serum of A, give no agglutination. Cells of B added to the serum of B give no agglutination. But cells of B agglutinate in the serum of A, and, reciprocally, cells of A agglutinate in B serum. All this can be explained by assuming that A individuals have an A antigen on their red cells and have antibodies to B in their serum. (They have no A antibodies in their serum, since that would be suicidal.) Similarly, B individuals have B antigen on their red cells and antibodies to A in their serum. Transfusion of A to B or B to A, therefore, will give agglutination.

Blood types fall into four major groups with respect to the A, B antigens; A, B, AB, and O. As shown in Table 9-2, A carries the A antigen and B antibodies; B carries the B antigen and the A antibodies; AB carries both the A and B antigens on the same cell and no antibodies, while O carries

Table 9-2

HUMAN BLOOD GROUPS

Genotypes	Cell antigen	Serum antibodies	Blood group
$L^A L^A$ or $L^A l$	A	Anti-B	A
$L^B L^B$ or $L^B l$	B	Anti-A	B
$L^A L^B$	A and B	None	AB
ll	Neither A nor B	Anti-A and B	O

neither antigen and has both antibodies. The group to which an individual belongs is inherited as a single gene and the difference between an A and a B individual is an allelic difference.

The inheritance of blood type, then, corresponds to the inheritance of a gene that controls the formation of a specific antigen. The gene is called the L gene in honor of the great immunologist, Carl Landsteiner. Individuals of type A carry an allele L^A which directs the formation of antigen A. Individuals of type B carry a second allele L^B which directs the formation of antigen B. Individuals of type O carry a third allele, 1, and neither A nor B antigen is synthesized. We would expect to find genotypes representing all possible combinations of the three alleles. These combinations consist of the six different genotypes listed in Table 9-2. The six genotypes give four phenotypes, A, B, AB, and O, since L^A and L^B are dominant to 1 but show no dominance with respect to each other.

The genetic consequences of this type of inheritance are clear. An individual of blood group O might arise from the mating of two individuals of blood group A (i.e., from AO $\times$ AO) or from two individuals, one A and the other B (AO $\times$ BO), but an individual of blood group O could not have an AB parent. Conversely, an AB individual could not possibly arise from a mating of two individuals of blood group O. For this reason, blood grouping is of critical importance to legal medicine in helping solve such thorny cases as issues of disputed parentage. Although the inheritance of the A and B antigens is reasonably well understood, the inquiring student may wonder how the L^A gene, while determining the formation of a specific antigen A, at the same time results in the formation of antibodies against B. The best information is that the association of L^A and anti-B antibodies is accidental, but the problem is still open. It must be emphasized that other blood group systems in man are *not* associated with the simultaneous presence in the serum of antibody to the antigenic product of the alternate allele.

Other blood traits in man are also known to be inherited as an allelic series. The Rh factor was discovered by testing human red cells with antiserum produced against red cells of the Rhesus monkey. Some human bloods reacted with these antiserums, others did not, revealing that an antigen similar to one on the monkey cells was present on some human blood cells but not on others. The antigen was called the Rhesus or Rh factor and it has proved of great interest since an association has been established between the Rh factor and hemolytic anemia of the newborn. Genetic investigations have revealed many different Rh alleles, and the inheritance of the Rh factor represents a multiple allelic series or a series of very closely linked genes. Although the Rh factor is an inherited trait, its relation to hemolytic anemia of the newborn is an immunological problem.

In matings of an Rh^+ father and an Rh^- mother, a situation can arise in which the Rh^- mother carries an Rh^+ embryo. You can readily figure out the genotypes of the two parents required for this situation. In some cases, the Rh factor (antigen) of the embryo can cross the placenta and induce antibody formation in the mother. (Since she is Rh^-, she will react to the Rh factor as she would to any other foreign antigen.) We thus have a situation in which the Rh factor of the embryo induces antibodies in the mother, and these antibodies can now return and damage the embryo by agglutinating and destroying the embryo red cells (Fig. 9-4). Since the total amount of Rh factor in the embryo is small, the total amount of antibody formed in the mother will be limited and will have relatively little effect on the first born. With succeeding Rh+ pregnancies and continued immunization, the level of anti-Rh antibodies will increase in the mother's serum, increasing amounts of antibodies can enter the fetal circulation, and this in time can give rise to a severe anemia in the fetus. After a few pregnancies, antibody levels can become dangerously high, and the newborn may require a complete exchange transfusion to prevent death. It is clear that knowledge of Rh compatibility in prospective parents is of vital importance. We can then be alert for the possible appearance of hemolytic anemia in the newborn child. If it does exist, a complete exchange transfusion will save his life.

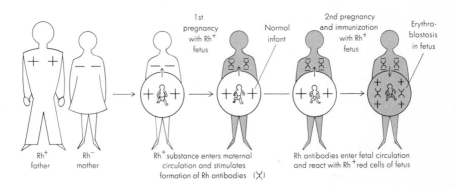

Fig. 9-4. Rh-factor incompatibility (after Srb and Owen).

Still other antigenic differences are known for blood, and even in the ABO group, new differences are still being uncovered. In fact, human blood constitutes a magnificent laboratory for genetic experimentation. We have discussed the brilliant investigations of human hemoglobin and the inheritance of blood types and the Rh factor. In the years ahead, it may well be possible that use can be made of circulating blood cells *in vivo* to study problems of cell populations in a manner similar to that in which we study bacterial genetics.

Up to now, our discussion of genetics has been confined mainly to the transmission of single genes and their alleles and to the relationship of single genes to specific biochemical reactions. The inheritance of many characteristics, however, appears to be governed by more than one gene.

These include such traits as susceptibility to certain diseases, height, the development of the nervous system, skin color, etc. An extended analysis of multigenic inheritance is beyond the intention of this book; but, because of its importance to human genetics, we will briefly describe one case.

SKIN COLOR

Let us see what would happen if two unlinked genes, each contributing equally, were to determine the difference between black skin color and white. Let the capital letters Y_1 and Y_2 stand for the black genes, and y_1, y_2 for the white alleles. Assume white and black are homozygous for the two genes. A cross between black Y_1Y_1,Y_2Y_2 and white y_1y_1,y_2y_2 would give the F_1, Y_1y_1,Y_2y_2, intermediate in color between the parents. A cross between F_1, Y_1y_1,Y_2y_2, and white, y_1y_1,y_2y_2, would give three genotypes and three phenotypes: yy,yy = white, $Y_1y_1Y_2y_2$ = the same as F_1, Yy,yy = intermediate between F_1 and white. The three phenotypes would be in the proportion 1:1:2. (We have left out the subscripts in two genotypes since the *total number* of genes is the determining factor, not whether it is an allele of the first or the second gene.) If you work out an $F_1 \times F_1$ cross, you will find the results to be 1 white, 1 black, and 14 with varying amounts of intermediate pigmentation.

In human skin pigmentation, the genetic difference between yellow, black, red, and white includes many genes. Estimates vary for the total number involved, but the evidence points to at least eight and probably many more. Assume the number is ten, that they are unlinked, and the conditions stated above pertain. In a mating between two people, both heterozygotic for every gene, 1 in about 1,000,000 of their children would be either white or black, a rare occurrence; the other children would be intermediate. On the other hand, a mating between white and intermediate could produce no children darker than either parent.

In this curtailed analysis of multigenic inheritance, certain simplifying assumptions were made. Since a large number of genes govern the inheritance of skin color, we would expect some of them to be linked and thus not able to make equal contributions. The conclusions we arrived at, however, will not be badly distorted. This is affirmed by analysis of multigenic traits in plants and animals under conditions of controlled breeding.

In this chapter, we have attempted to present an outline of the principles of human genetics. There are other aspects of human genetics, each with its unique contribution to the diversity of genetic mechanisms in man, that we cannot describe because of space limitations—for example, the transmission of color blindness and hemophilia, which follows a pattern of "sex-linked" inheritance because the genes responsible for these traits are on the X-chromosome but not on the Y. (Note that XX = female is normally diploid for a sex-linked trait, but XY = male is haploid.) The significant thing is that the genetic material of man, in its transmission and action, is similar to that found in other organisms. We are thus increasingly sure that information obtained by study of microorganisms, insects, plants, and animals will be pertinent to the solution of the hereditary problems of man. Perhaps of greatest interest is our present conviction that human genetics itself will make important contributions to the basic problems of genetics, a statement that would have been considered implausible ten years ago.

Heredity
and Hiroshima

CHAPTER TEN

In the course of the past 60 years, we have learned much about heredity and its underlying chemical basis. In turn, genetic knowledge has contributed to man's welfare on many levels. The material rewards to society derived from such knowledge have already been large, and will undoubtedly prove larger in the future. An exciting chapter in genetic history has been written in the production of high-yield strains of hybrid corn, of high-yield wheat strains resistant to attack by the fungal parasites, rust and smut, which formerly caused nationwide crop failure, and in the more recent, slightly incredible, use of X-ray sterilized male blow flies to help eradicate the pestilent screw worm in the South. These all attest to the value of genetic knowledge in eliminating pestilence and in breeding animals, plants, and microorganisms to give hardier, more productive strains and to increase the world's available resources of raw materials, antibiotics, and food.

Today a glance at the newspapers, with articles on population explosions, the drive of newly emerging nations for economic self-sufficiency, and live polio vaccines obtained from mutant virus strains, will again remind you of some of the consequences of man's curiosity about the so-called ivory-tower concepts of the gene, DNA, and messenger RNA.

However, let us turn our attention to still another topic of contemporary importance, caught up in recurrent ominous headlines, to a topic that now engages the attention of people throughout the world—the global genetic problem arising from the harnessing and exploitation of atomic energy. We know that a portion of the stupendous energy released from nuclear fission and fusion is released in the form of high-energy ionizing radiations, and in the form of radioactive isotopes of elements such as strontium and carbon. Ionizing radiations are highly mutagenic. Before the advent of atomic energy, the known mutagenic agents with which the human population had to contend were not of alarming concern. For example, ultra-violet light has low tissue-penetrating properties and though it causes burning of skin and may even induce skin cancer, it has no marked mutagenic effect on other human cells. Chemical agents such as mustard gas are deleterious only when fools deliberately set them loose to be inhaled, and base analogues are ordinarily not present in the diet, although they may be used to treat or control various cancers. In general, these are mutagenic agents under our control. They can be used or forgotten at will.

Ionizing radiation, however, is a different kettle of fish. High-energy radiation is capable of deep tissue penetration and is massively destructive and mutagenic. Since our environment is bombarded by a certain amount of background radiation from naturally occurring radioactive elements and from outer space, ionizing radiation is a mutagen we cannot entirely control and against which we are never completely protected. Our present concern is with the increasing amounts spewn into our environment from nuclear explosions. Each of us now receives more radiation than did our forefathers, and our children may receive more. Since the background radiation has shot upward in the past 20 years and since ionizing radiation is mutagenic, we must give thoughtful consideration to the possible genetic effects of this increase and of how great an exposure the human population can withstand.

Our concern with ionizing radiation is of two kinds, its effects on germinal tissue and succeeding generations, and its effects on somatic tissue, on us. Let us first consider somatic effects. Is there an association with somatic abnormalities, such as cancer, and radiation? The frequency of leukemia, a cancer characterized by an excess formation of white blood cells, can be increased by radiation. In the population of Hiroshima and Nagasaki that survived the atom bombs, there was a marked increase in leukemia during the subsequent five years. The association between leukemia and radiation is exhibited in other groups that have somehow been exposed to increased radiation. There is therefore a clear relationship between increasing amounts of radiation and increasing incidence of cancer.

We know that radiation results in somatic mutation. In view of the correlation found between leukemia and radiation, does this mean that cancer is mutational in origin? We do not know. We have experimental evidence that some animal leukemias can be induced by a virus; further, we know that interactions between bacteria and bacterial viruses are modified by radiation. We cannot, therefore, conclude that the increase in leukemia associated with an increase in radiation arises as a consequence of mutation, for it could result from the interaction of irradiation, virus, and host. This is, perhaps, a fair assessment of the relation of radiation to cancer in general. Radiation could induce cancers for a number of different reasons. Since cancers have diverse origins, they may be caused by a virus, a mutation, or other factors. Regardless of the details, however, there is an association between an increasing amount of radiation and an increasing incidence of cancer. Radiation is clearly a potential somatic hazard.

We know of still other somatic effects of radiation, for a relationship between irradiation and aging has been observed. With increasing amounts of radiation, longevity is shortened. In fact, from animal experiments it appears that the shortening of human life may well be measured in days per unit dose. Again, increasing radiation must be viewed as a somatic hazard.

Uncontrolled release of nuclear energy also imperils the human population because it increases the abundance of certain radioactive elements. For instance, radioactive strontium 90 formed as a consequence of hydrogen-bomb explosions is extremely dangerous, since it has a long half life and can replace calcium in bone. It is carried into the ionosphere and is ultimately washed down and widely deposited over the earth's surface by rain and snow. From the soil, strontium 90 is absorbed by growing plants, the plants are ingested by cattle, and strontium 90 appears in milk and milk products. When taken into the body, this radioactive isotope can take the place of calcium in the bone tissue, particularly in growing children. Strontium 90 in bone constitutes small centers of radioactivity and, if present in sufficient amounts, can give rise to bone cancer and a variety of other abnormalities. This is truly a new cycle of nature, a "twentieth-century cycle." The amount of strontium 90 has already been markedly increased by nuclear explosions, and we have no clear-cut answers as to how much strontium 90 can be safely tolerated in the human diet or how big an increase in the total amount of strontium 90 can be absorbed without adverse effect on the population as a whole.

Safety levels in man are difficult to determine. They require long-term study, but our restless world denies us time. Decisions affecting the use of nuclear energy must be made, but, unfortunately, our information about the deleterious effect in man per unit increase of radiation is inadequate to permit us to offer unequivocal biological advice. The thoughtful person must agree that an increase in radiation is attendant with hazard, and that any political decision leading to an increase in radiation must be made on the basis of the gravest, most soul-searching considerations.

A second problem posed by bomb testing is the effect of ionizing radiation on germinal tissue. There is no question that mutation rates are enhanced by radiation. This was shown years ago by H. J. Muller with X-rays and Drosophila. By far the great majority of such mutations are lethal or deleterious. As background radiation increases, the average dose received by each of us increases, and presumably the total number of mutations present in our gametic cells also increases. If the total number of transmissible mutant genes is on the increase, the number of genes having a deleterious developmental effect obviously must also rise. For the sake of future generations, we desperately need information on how big an increase in the genetic burden a human population can withstand and survive!

In the populations of the cities of Hiroshima and Nagasaki that survived the atom bomb, despite the observed increase in frequency of leukemia, no substantial rise in the frequency of spontaneous abortions or in the number of stillborn infants has been noted. These findings, though fragmentary, are of interest, since the rate of spontaneous abortions and stillbirths perhaps should provide some clue about the increase in dominant mutant lethal genes in the irradiated population, and the increase in dominant lethals in turn might give at least a rough estimate of the increase in recessive lethal genes. But no increase was found. Perhaps dominant lethals in humans are rare compared to recessive lethals, or perhaps dominant lethals are quickly filtered out by cell death. Inbreeding experiments are necessary to answer this point, but such experiments are denied us.

Many indirect methods have been used to estimate the so-called "genetic load" (number of recessive lethal and deleterious genes carried in the human population). Analysis of the vital statistics of various small populations in which consanguineous marriages have occurred for many generations permits a rough estimate of genetic load. The estimates have varied, but in general they have indicated a smaller load than would have been anticipated from studies of other organisms. For example, relatively large numbers of recessive lethals can accumulate and be maintained in laboratory populations of Drosophila, with a consequent reduction in the vitality of the population when inbred.

In human populations, the number of recessive lethals appears to be relatively small. It must be emphasized that the data on human populations are few and the analysis is uncertain. However, it is possible that lethals do not accumulate in man, but are eliminated within the first few divisions of the fertilized egg. If this is true and if the data obtained from the Nagasaki and Hiroshima populations illustrate what we can expect in human populations, then the evidence suggests that an increase in radiation may constitute a greater somatic than a germinal hazard; but irradiation and its effect on the genetic load of the human population remain virtual unknowns.

One fact does remain crystal clear. Increased radiation represents a serious human hazard, a hazard of recent origin and one that must be studied in detail. In the absence of overriding political considerations, background radiation obviously should not be increased without a sounder knowledge of the biological consequences of such an increase. How can more detailed information about the effects of radiation on human populations be obtained? The answer, in part, must come from population studies of other organisms, and it also hinges on how successful we are in securing detailed information about the genetics of man. As mentioned earlier, techniques are now being developed that in time may enable us to grow differentiated cells in tissue culture, and we may soon be able to study differentiated human cells much as we now grow and study bacterial cells. Perhaps the methods that have proven fruitful in defining genetic mechanisms in microorganisms will then prove fruitful in defining genetic problems in man, including the effects of radiation. The problems are immediate and urgent and require the thoughtful attention of all of us.

All of heredity cannot be presented in this slim volume. We have attempted to present the current ideas on the chemical basis of heredity and its mechanisms of action in the living cell. In addition, we have described some of the latest trends in genetic research, the problems that confront us, and the expectations that stimulate us. To do this, we have touched but lightly on the contributions of the giants of the past. And we have omitted, or merely hinted at, many other fields of genetic inquiry that are bustling and productive. If we have succeeded in whetting your appetite, we leave it to your curiosity to explore more deeply into the science of genetics. In our own biased opinion, we think that the future of genetics can be simply stated by these lines from a poem by e.e. cummings:

—listen: there's a hell
of a good universe next door; let's go

SELECTED READINGS

There are many excellent texts of general genetics. We refer to but three of these and one collection of epic papers.

Sager, R., and Francis J. Ryan, *Cell Heredity*. New York: Wiley, 1961.

Sinnot, E. W., L. C. Dunn, and Th. Dobzhansky, *Principles of Genetics,* 5th ed. New York: McGraw-Hill, 1958.

Srb, A. M., and Ray D. Owen, *General Genetics*. San Francisco: Freeman, 1952.

Peters, James A., ed., *Classic Papers in Genetics*. Englewood Cliffs, N. J.: Prentice-Hall, 1959.

CHAPTER ONE

Swanson, C. P., *Cytology and Cytogenetics*. Englewood Cliffs, N. J.: Prentice-Hall, 1957.

Mazia, Daniel, "How Cells Divide," *Scientific American,* 206 (September, 1961), 100.

CHAPTER TWO

Hotchkiss, Rollin D., and Esther Weiss, "Transformed Bacteria," *Scientific American,* 195 (November, 1956), 48–53.

Crick, F. H. C., "The Structure of the Hereditary Material," *Scientific American,* (October, 1954).

CHAPTER THREE

Beadle, G. W., *Genes and the Chemistry of the Organisms,* Science in Progress, 5th series. New Haven: Yale University Press, 1947.

Wagner, R. P., and H. K. Mitchell, *Genetics and Metabolism*. New York: Wiley, 1955.

CHAPTER FOUR

Anfinsen, C. B., Jr., *The Molecular Basis of Evolution*. New York: Wiley, 1959.

CHAPTER FIVE

Vogel, H. J., V. Bryson, and J. O. Lampen, ed., *Informational Macromolecules*. New York: Academic Press, 1963.

Meselson, M., and F. W. Stahl, "The Replication of DNA in *Escherichia coli,*" *Proceedings of the National Academy of Sciences,* Washington, 38 (1958), 953.

Neirenberg, Marshall W., "The Genetic Code: II," *Scientific American,* 208 (March, 1963), 80.

CHAPTER SIX

Bonner, D. M., Y. Suyama, and J. A. DeMoss, "Genetic Fine Structure and Enzyme Formation," *Federation Proceedings,* 19 (1960), 926.

Benzer, S., "The Fine Structure of the Gene," *Scientific American,* (January, 1962).

CHAPTER SEVEN

Beale, G. H., *The Genetics of Paramecium aurelia.* New York: Cambridge University Press, 1954.

Moyed, H. S., and H. Edwin Umbarger, "Regulation of Biosynthetic Pathways," *Physiological Reviews,* 42 (1962), 444.

CHAPTER EIGHT

Wollman, Elie, and François Jacob, "Sexuality in Bacteria," *Scientific American,* 195 (July, 1956), 109.

Zinder, Norton O., "Transduction in Bacteria," *Scientific American,* 199 (November, 1958), 38.

Jacob, François, P. Schaeffer, and Elie Wollman, "Episomic Elements in Bacteria," *Microbial Genetics,* 10th Symposium of the Society for General Microbiology. New York: Cambridge University Press, 1960.

CHAPTER NINE

Harris, H., *Human Biochemical Genetics.* New York: Cambridge University Press, 1960.

CHAPTER TEN

Beadle, George W., "Ionizing Radiation and the Citizen," *Scientific American,* 201 (September, 1959), 219.

Knipling, Edward F., "The Eradication of the Screw-worm Fly," *Scientific American,* 203 (October, 1960), 54.

Index